QUALITY CONTROL IN CONSTRUCTION

BEST PRACTIC

AVINASH GUPTA

BE (civil), PG Construction Management, MIE, C. Egg

QUALITY CONTROL COVERS PLANNING EXECUTION STANDARDS

Quality Control in Construction
Best Practices

Author: Avinash Gupta

First published: JULY 2025

Published By: Author Self Published

ISBN : 978-93-343-2822-6

Book Overview

This book serves as an in-depth guide to quality control in construction, focusing on essential principles, methodologies, and best practices for ensuring excellence in luxury residential, office, and commercial building projects. Drawing on over 30 years of experience in the field, the author aims to equip construction professionals with the knowledge and tools necessary to implement effective quality control measures that enhance project outcomes, safeguard investments, and elevate client satisfaction.

CONTENTS

PREFACE

In the changing world of construction, the importance of quality control is very high. As projects become more complicated, it is crucial to ensure that every part meets high standards. This is important not only for making structures last longer, but also for the safety and satisfaction of everyone involved. The book "Quality Control in Construction: Best Practices" is a complete guide that highlights key points of quality management throughout the construction process.

This book is organised to give both theoretical knowledge and practical applications. It starts by discussing why quality control matters and how it is essential for successful construction projects. We explore various standards and rules that influence quality practices, emphasising their role in ensuring safety and compliance.

Each chapter builds on the previous one, offering specific strategies for effective project planning, choosing contractors, and managing construction sites. We stress the need for good workmanship and the testing and inspection methods needed to maintain these quality standards. Special focus is given to structural quality control and the unique challenges that come with mechanical, electrical, and plumbing (MEP) systems.

Sustainability is another important theme in this book, reflecting the increasing concern for the environment in the construction industry. We discuss how to include sustainable practices in quality control measures, showing that responsible construction goes hand-in-hand with achieving high quality.

We also highlight the importance of documentation and thorough reporting, as these practices are crucial for transparency and accountability. Additionally, we cover the need for regular audits and assessments to ensure continuous improvement and adherence to best practices.

"Quality Control in Construction: Best Practices" is more than just a set of guidelines; it aims to promote a culture of quality and excellence in construction. By sharing lessons learned from past projects and applying the principles of Total Quality Management, we hope to equip professionals with the tools they need to improve their quality control efforts.

As you read this book and explore the world of construction quality control, we hope you find inspiration and practical knowledge that helps you contribute to building safer, stronger, and more sustainable structures. Together, let us create a future focused on quality.

CHAPTER 1

THE SIGNIFICANCE OF QUALITY CONTROL

"Quality control (QC) in construction is a systematic and disciplined process designed to ensure that all construction activities, materials, and outcomes meet predefined standards, regulatory requirements, and client expectations."

Definition and Scope of Quality Control

Quality control (QC) in construction includes a wide range of activities, such as inspections, testing, documentation, and audits, that need to be done at every stage of a project. QC covers not just the quality of materials and workmanship, but also the management of design, planning, execution, and the final completion of the project. Every stage of construction is important, and strong QC practices help ensure that each milestone meets or exceeds the expected standards.

In luxury construction projects, where both appearance and functionality are very essential, the role of QC becomes even more crucial. The high level of detail and precision required for luxury projects means that strict quality checks need to start right from the beginning. Whether it's high-end residential properties, commercial buildings, or custom interiors, ensuring quality must be a fundamental part of the project from the start.

Expanded Scope of Quality Control

Design Phase Quality Control

The design phase of quality control (QC) is all about making sure that project plans are correct, clear, and doable. This involves checking all design documents like drawings and specifications very carefully. Critical activities during this phase include having meetings with the team to review the designs and ensure they meet necessary rules and standards. Value engineering helps find ways to save costs without reducing quality. Risk assessments are done to spot any potential issues that could affect the project's safety or feasibility. Getting input from all team members ensures that practical concerns are addressed early. By focusing on quality in the design phase, organisations can greatly reduce mistakes and rework later on.

Pre-Construction Quality Control

Pre-construction QC lays the groundwork for a successful project by setting clear quality goals and confirming that everyone is ready. This phase involves defining specific quality targets that match the project's goals, such as performance standards or sustainability aims. It's essential to check that all materials meet required specifications, including verifying certifications. Assessing the contractor's skills and experience ensures they can deliver high-quality work, which might involve looking at past projects and getting references. Meetings are held with all stakeholders to explain quality expectations and project requirements, ensuring everyone understands their roles. By building a strong QC foundation in this phase, risks can be reduced and expectations clarified for everyone involved.

Construction Phase Quality Control

During the construction phase, QC is crucial for making sure that the work done meets quality standards. Regular inspections are carried out throughout construction to check the quality of work against the design specifications. This includes examining aspects like safety and finish quality. Materials used in construction are also tested to ensure they are suitable and meet required standards. Assessing the workmanship of contractors and subcontractors is essential, involving visual checks and evaluations based on quality criteria. Monitoring progress against project milestones ensures that work is completed on time and as planned. By managing quality actively during this phase, organisations can quickly identify and resolve the problems, reducing rework and staying on schedule.

Post-Construction Quality Control

Post-construction QC is important for ensuring that the completed project meets all quality standards before handing it over to the client. This phase includes conducting final inspections to check the overall quality of the project and making sure all systems work properly. The handover process involves providing the client with all necessary documents, including warranties and maintenance guides, for future use. Reviewing warranties ensures that the contractor stands by their work, and it's essential to inform clients about how to use these warranties effectively. Collecting feedback from clients and stakeholders helps identify areas for improvement, creating a loop that is vital for enhancing future QC processes. By implementing strong quality control measures in the post-construction phase, organisations can ensure long-term satisfaction and performance, showing their commitment to quality even after the project is finished.

Advantages of Rigorous Quality Control:

Implementing a rigorous quality control process provides tangible benefits that extend beyond meeting construction standards. By emphasising preventative measures and thorough monitoring, construction firms can secure their project outcomes and enhance overall operational performance.

Enhanced Reliability and Safety:

A well-implemented QC system reduces the chances of defects and failures, ensuring that the final structure is aesthetically pleasing and structurally sound. This is particularly vital in high-stakes projects where safety cannot be compromised. Luxury constructions, where both form and function are paramount, demand unparalleled attention to detail in terms of material selection and craftsmanship. A robust QC system is essential for ensuring reliability, which protects both the integrity of the structure and the safety of its occupants.

Cost Savings Through Prevention:

While implementing quality control processes requires an upfront investment, these measures are significantly more cost-effective than corrective actions that need to be taken after the project is completed. When issues are caught early, there's a lower likelihood of costly rework, project delays, or penalties due to non-compliance. In a worst-case scenario, substandard quality could result in legal claims or structural failure. Early detection through QC saves money by minimising such risks and ensuring that the project stays within budget.

Boosting Client Satisfaction and Loyalty:

Satisfied clients are the lifeblood of a successful construction business. By consistently delivering high-quality outcomes, construction firms can develop a reputation for excellence, leading to repeat business, referrals, and positive industry recognition. In the luxury construction sector, client expectations are extremely high. Adopting a thorough QC process helps meet and exceed these expectations, cementing relationships with high-end clients and positioning the firm as a leader in its field.

Regulatory Compliance and Reduced Legal Risk:

Stringent QC practices help ensure that a construction project complies with all relevant local and national regulations. Failure to comply with these standards can result in fines, project delays, or legal action, all of which can severely impact a construction firm's reputation and financial standing. By maintaining high standards of quality control, firms are better equipped to meet regulatory requirements, protecting themselves from legal repercussions and enhancing their overall industry standing.

Reputation as a Quality Leader:

A commitment to quality differentiates a construction firm from its competitors. When a company becomes known for delivering high-quality work consistently, it is easier to attract high-end clients, top-tier employees, and industry partnerships. In an industry where reputation is everything, a strong QC process is an invaluable asset for long-term success. Consequences of Inadequate Quality Measure While the benefits of a strong QC system are clear, the consequences of poor or inadequate quality control can be severe. A failure to implement effective QC measures can result in critical safety issues, increased project costs, and significant damage to a company's reputation.

Structural Failures and Safety Risks:

One of the most serious risks of insufficient quality control is the potential for structural failures, which can endanger both the building's occupants and the construction workers on site. Without proper monitoring and testing, materials may not meet the necessary strength requirements, or workmanship might not adhere to safety standards. This can lead to catastrophic failures, such as building collapses, electrical fires, or plumbing issues. These failures pose a serious risk to human life and result in costly legal liabilities and penalties for the construction firm.

Increased Costs and Project Delays: Poor quality control often leads to defects that must be addressed after the construction has progressed or even after the project is completed. These defects can be expensive to fix and may lead to significant delays in project timelines. Unplanned rework, increased labor costs, and material waste all contribute to exceeding project budgets and eroding profit margins. In the luxury sector, where the margin for error is especially narrow, these financial implications can be particularly devastating.

Reputation Damage: The construction industry is highly competitive, and reputation is critical to securing new business. Word-of-mouth and client recommendations play a significant role, particularly in luxury construction. Poor quality control, leading to unsatisfactory results, can result in negative client feedback, poor reviews, and a diminished standing within the industry. Over time, this can erode a company's ability to secure new projects and retain high-end clients, severely impacting its bottom line.

Checklist for Understanding Quality Control:

This checklist helps project managers and engineers assess their understanding and approach to quality control.

Define quality control objectives for the project: Ensure these objectives align with both the client's expectations and industry standards.

Identify key stakeholders involved in quality assurance:

Stakeholders might include engineers, architects, project managers, and external auditors. It's critical to assign clear responsibilities to each group.

Establish metrics for measuring quality success: These metrics should be quantifiable and regularly reviewed throughout the project.

Incorporate routine inspections and audits: Conduct regular quality control checks at every phase of the project.

Train team members on QC processes: Quality control should be a shared responsibility, and all workers and engineers must be adequately trained on the necessary QC standards and protocols.

In-House Audit Methods for Quality Assurance

In-house audits are critical for maintaining and improving quality standards within an organisation. They provide a structured method for assessing processes, identifying inefficiencies, and ensuring compliance with established quality benchmarks.

Pre-Audit Planning : The audit process begins with pre-audit planning, which involves defining the audit scope, objectives, and criteria. Establishing a clear scope is essential; it outlines the specific areas, departments, and processes to be reviewed, as well as the timeframe for the audit. The objectives should align with the organisation's quality goals, such as evaluating compliance with internal policies and identifying areas for process improvement. Additionally, establishing criteria based on regulatory requirements, industry standards, and internal policies serves as benchmarks against which processes will be assessed. It is also important to assemble an audit team with the right mix of experience and expertise to ensure objectivity.

Audit Execution: During the audit execution phase, the team conducts field observations, engages with personnel through interviews, and reviews relevant documentation to gain a comprehensive understanding of current practices. Direct observation of construction activities and material handling helps assess adherence to safety protocols and quality control measures. Engaging with team members across various levels provides insights into their understanding of quality standards and any challenges they face. Utilising standardised checklists and auditing tools ensures thorough coverage of all necessary areas.

Post-Audit Review: After the audit, a post-audit review is essential for documenting findings. This report should summarise areas of compliance and non-compliance, classify non-conformities, and develop corrective action plans. Each identified issue should be addressed with immediate actions for urgent corrections and long-term strategies to prevent recurrence. Scheduling follow-up reviews to assess the implementation of corrective actions and establishing metrics for evaluating effectiveness are crucial for maintaining compliance.

Ultimately, integrating audit findings into the organisation's quality management framework fosters a culture of continuous improvement. This ensures that the organisation remains committed to excellence in quality assurance, reinforcing its dedication to maintaining high standards and optimising operational efficiency.

Kaizen Practice in Quality Control

The Kaizen philosophy, which means continuous improvement, is very useful in quality control. This approach encourages everyone in the team to constantly look for ways to improve processes and practices. By adopting Kaizen practices, organisations can create a culture of quality that involves everyone.

Kaizen Workshops: Regular workshops are an effective way to bring workers together to discuss areas where improvements can be made. These workshops provide a platform for team members to share their experiences and propose solutions to problems they encounter on the job. This collaborative environment helps generate new ideas and encourages a sense of ownership among workers.

Feedback Loops: Creating feedback mechanisms is essential for empowering workers. By providing ways for employees to voice their concerns and suggest improvements, organisations can enhance quality control measures. This could be through suggestion boxes, regular meetings, or digital platforms where team members can share their thoughts. When workers feel their input is valued, they are more likely to contribute positively to the quality control process.

Training Programs for Engineers and Workers

Ongoing training is crucial for maintaining high standards in quality control. By offering regular workshops, certifications, and on-site training, organisations ensure that everyone— from labourers to project managers—understands and aligns with quality control standards.

Quality Control Training Modules: Training should cover important topics such as material inspection, safety protocols, and defect prevention. These modules help all team members understand what to look for in materials and how to ensure that safety measures are followed, which is essential for preventing issues before they arise.

Workmanship Standards Training: On-site training focused on critical skills like masonry, carpentry, plumbing, and electrical work is vital. This hands-on approach allows workers to learn best practices directly related to their tasks, helping them produce higher-quality work.

Leadership Training for Supervisors: It is also important to develop leadership skills among site managers and supervisors. Effective leadership ensures that quality control is monitored and maintained on-site. Training for supervisors can include topics like communication skills, team management, and techniques for overseeing quality control processes.

By incorporating Kaizen practices and investing in training, organisations can foster a culture of continuous improvement and ensure that quality control remains a top priority throughout the construction process. This commitment to quality ultimately leads to better results and greater satisfaction for all stakeholders involved.

"Quality control in construction is not optional — it's essential. It ensures safety, durability, and compliance with standards. Without it, structures become liabilities, not assets. Every inspection protects both lives and investments."

■■

CHAPTER 2

STANDARDS AND REGULATIONS IN QUALITY CONTROL

Overview of Relevant Laws and Regulations

Construction quality control is based on a set of laws and regulations that ensure safety, structural strength, environmental sustainability, and the overall quality of buildings. These rules exist at local, national, and international levels and cover everything from design and material selection to environmental impact and worker safety. Understanding these regulations is crucial for reducing risks and ensuring successful project outcomes, as it helps construction teams make informed decisions that protect workers and meet quality standards, ultimately leading to safe and reliable buildings. Key regulations may include:

Local Building Codes: These codes and regulations are set by municipal or regional governments that define the minimum standards for the design, construction, alteration, and maintenance of buildings. These codes vary by location and may include rules for zoning, fire safety, plumbing, electrical work, and accessibility.

Occupational Safety and Health Administration (OSHA) Standards: In many regions, safety regulations like OSHA ensure the protection of workers on construction sites. These guidelines cover equipment handling, fall prevention, electrical safety, and how to communicate hazards.

Environmental Regulations: These laws manage the environmental impact of construction activities, including waste management, emissions control, and the use of sustainable resources. Following these regulations helps projects achieve environmental goals and avoid penalties.

Sticking to these regulations not only ensures compliance with the law but also boosts the project's credibility and reduces the risk of costly penalties or delays.

Key Industry Standards

Several globally recognised industry standards are crucial for quality control in construction. These standards set best practices for material selection, construction methods, safety protocols, and overall project management:

ISO 9001: This is one of the most widely adopted quality management standards in various industries, including construction. It focuses on a process-driven approach, emphasising continuous improvement, customer satisfaction, and regular internal audits. Companies that follow ISO 9001 must keep strict documentation and maintain a feedback loop. This helps improve efficiency, product quality, and competitiveness in the market.

ASTM International Standards: ASTM (American Society for Testing and Materials) provides over 12,000 standards for various of materials, products, and services. These standards ensure that construction materials, like concrete and steel, meet safety and durability requirements. Following ASTM standards gives construction firms a reliable framework for procuring, testing, and ensuring the quality of materials.

Building Information Modelling (BIM) Standards: The use of BIM has become important recently. BIM standards ensure that digital models of a building's physical and functional characteristics meet quality and accuracy criteria. Using BIM improves coordination among all stakeholders, enhances decision-making, and reduces design flaws during construction.

Importance of Compliance

Maintaining compliance with laws, regulations, and industry standards is essential for a construction firm's success. Beyond the legal requirement, compliance serves as a testament to the firm's commitment to safety, quality, and ethical practices. Adhering to regulations and standards ensures several benefits:

Risk Mitigation: Adhering to safety, environmental, and quality standards significantly reduces the likelihood of accidents, defects, and legal disputes. By following established guidelines, construction firms can create safer work environments that protect both workers and the public. This proactive approach helps identify and manage potential hazards before they lead to serious issues, thus minimising risks associated with construction projects.

Enhanced Reputation and Credibility: Companies that consistently meet or exceed regulatory and industry standards are viewed as reliable and trustworthy by clients and stakeholders. A strong compliance record can enhance a firm's reputation, making it more appealing to potential clients. This credibility can be a significant competitive advantage when bidding on projects or forming partnerships, as clients are more likely to choose firms known for their commitment to quality and safety.

Market Competitiveness: In today's construction market, clients increasingly prefer to work with firms that demonstrate a strong commitment to quality and compliance. Meeting global standards like ISO 9001 or ASTM boosts a company's credibility and improves its chances of securing high-profile projects. Additionally, firms that are compliant are often better positioned to respond to client requests for proposals (RFPs), which frequently require evidence of adherence to specific standards.

Cost Efficiency: Non-compliance can lead to significant financial penalties, project delays, or the need for redesigns, all of which can strain a company's resources. By investing in compliance from the start of a project, firms can avoid costly rework and project overruns. Furthermore, compliant practices frequently lead to more efficient operations, as they encourage the use of standardised processes that streamline work and reduce waste.

Improved Employee Morale and Retention: A strong focus on compliance fosters a culture of safety and quality within the organisation. When employees see that their employer prioritises their safety and adheres to regulations, it can boost morale and job satisfaction. This positive work environment can lead to higher employee retention rates, reducing turnover costs and maintaining a skilled workforce.

Sustainable Practices: Compliance with environmental regulations helps firms adopt sustainable construction practices. By following these laws, companies can minimise their environmental impact, manage waste effectively, and use resources more efficiently. This commitment meets legal requirements and appeals to clients and communities that value sustainability, further enhancing the firm's reputation.

Attracting Investment: Investors and stakeholders are more likely to support firms that demonstrate strong compliance records. A commitment to quality and safety can make a company more attractive to potential investors, who are often looking for businesses that minimise risks and maximise returns. Additionally, compliance can open doors to government contracts and funding opportunities that require strict adherence to regulations.

Checklist for Compliance

To ensure your construction project meets all necessary laws, regulations, and standards, it's important to follow a structured checklist. This checklist will help you stay organised and ensure that every aspect of your project aligns with compliance requirements.

Review All Relevant Local, National, and International Laws and Regulations: Start by thoroughly reviewing the laws and regulations that apply to your project. This includes local building codes, national safety standards, and any international regulations that may be relevant. In India, the **National Building Code (NBC)** is a key reference that outlines guidelines for building construction, safety measures, and design principles. Understanding these laws is essential for identifying what is required at each stage of construction. Make a list of the specific regulations that apply to your project and keep it accessible for the entire team.

Identify and Implement Applicable Industry Standards: Identify the relevant industry standards that should be integrated into your project. Important standards include **ISO 9001** for quality management, **ASTM and BIS codes** for material testing, and the **Bureau of Indian Standards (BIS)** codes. BIS codes cover various aspects of construction, including material specifications, testing methods, and safety protocols. Once identified, create a plan to implement these standards in your daily operations. This may involve training your team on these standards and ensuring that all materials and processes used in construction adhere to them.

Ensure All Project Team Members Understand Compliance Requirements

It's crucial that everyone involved in the project understands compliance requirements. Organise training sessions and workshops to educate your team about relevant laws and industry standards, including the NBC and BIS codes. Provide clear documentation that outlines compliance responsibilities for each role. This ensures that all team members, from labourers to project managers, are aware of their duties related to compliance and safety.

Conduct Regular Audits to Check Ongoing Compliance:

Regular audits are vital for ensuring that your project remains compliant throughout its lifecycle. Schedule audits at various stages of the construction process to assess adherence to laws and standards. These audits can be internal or conducted by third-party professionals. Use the findings from these audits to make necessary adjustments and improvements in your processes. This proactive approach helps identify issues before they escalate into significant problems.

Maintain Comprehensive Documentation for Regulatory Reviews and Audits:

Keeping detailed records is essential for compliance. Maintain comprehensive documentation that includes all permits, inspection reports, training records, and audit results. This documentation will not only help during internal reviews but will also be crucial for any regulatory inspections. Proper documentation serves as evidence that your project adheres to all necessary standards and regulations, including the NBC and BIS codes, and it can protect your firm in case of any legal or compliance-related issues.

By following this checklist, you can systematically ensure that your construction project meets all compliance requirements. This enhances the quality and safety of the project and builds a strong foundation for future work and client trust. Regularly revisit and update this checklist to reflect any changes in laws, standards, or project scope, ensuring that your compliance efforts remain effective and relevant.

"Standards and regulations in quality control ensure consistency and reliability, forming the foundation for trust and improvement."

■■

CHAPTER 3

QUALITY CONTROL IN PROJECT PLANNING

"Quality control in project planning is essential for preventing errors, ensuring standards are met, and delivering successful outcomes."

Quality control in project planning is a crucial aspect that ensures construction projects are carried out effectively, efficiently, and to the highest standards. It takes a proactive approach to minimise errors, comply with regulations, and achieve successful results. This process is vital for maintaining the integrity, safety, and aesthetic value of construction projects.

Establishing Quality Objectives

The first step in quality control is to set clear and measurable quality objectives. These objectives serve as guiding principles throughout the project, ensuring that all team members understand the expected standards.

Clarity and Specificity: Quality objectives need to be well-defined and specific. Vague goals can lead to misunderstandings among team members. For instance, instead of saying "use good materials," a more specific goal would be "use materials that comply with BIS standards for durability and safety." This clarity helps everyone know what is expected.

Alignment with Client Expectations: The objectives should reflect the client's needs and expectations. Engaging with clients during this phase ensures their vision is incorporated into the project. For example, if a client prefers eco-friendly materials, this preference should be included in the quality objectives.

Industry Standards and Regulations: Quality objectives must align with relevant industry standards, such as the National Building Code (NBC) and Bureau of Indian Standards (BIS) codes. This ensures compliance with legal requirements and best practices. For example, meeting NBC guidelines for structural safety can be a specific quality objective.

Key Aspects of Quality Objectives:

Structural Integrity: It's essential to ensure that the building meets safety and durability standards. Regular checks and adherence to guidelines are necessary. For example, using quality steel that meets IS 2062 standards ensures the structure can withstand loads.

Material Durability: Choosing materials that can resist environmental stress and wear over time is critical for the longevity of the building. This includes selecting weather-resistant paint or high-quality waterproofing membranes for foundations.

Aesthetic Quality: Maintaining high design and craftsmanship standards is especially important in luxury projects. For example, using handcrafted wooden fittings instead of mass-produced ones can elevate the project's aesthetic appeal.

Timely Project Completion: Completing the project on schedule without compromising quality is vital. This requires careful planning and resource allocation. For instance, setting a deadline for completing the roofing work before the monsoon season can help avoid delays.

Involvement of Key Stakeholders: Involving architects, engineers, contractors, and other stakeholders in setting quality objectives fosters a shared vision. Regular discussions and workshops facilitate collaboration and ensure everyone is aligned.

Ongoing Review and Adaptation: Quality objectives should be reviewed and adjusted regularly based on project developments and stakeholder feedback. For instance, if new safety regulations are introduced mid-project, the quality objectives should be updated accordingly.

Developing a Quality Management Plan

A comprehensive Quality Management Plan (QMP) is essential for guiding the project team in meeting established quality objectives. The QMP outlines specific processes, roles, and responsibilities, providing a clear framework for quality assurance and control.

Roles and Responsibilities: Clearly defining roles helps assign accountability. For example, the project manager oversees the overall quality control process, while a safety officer is responsible for ensuring compliance with safety regulations.

Expertise Matching: Assigning tasks based on expertise ensures effective quality control. For instance, a structural engineer should oversee structural integrity checks, while a plumbing expert should manage the installation of plumbing systems.

Quality Metrics

Establishment of Metrics: Create quantifiable metrics to measure quality at different project stages. This can include material specifications and safety compliance rates. For example, specifying that all concrete must have a minimum compressive strength as per IS 456.

Regular Monitoring: Implement a system for regular inspections and testing. For instance, conducting daily checks on the quality of incoming materials to ensure they meet the specified standards.

Communication Protocols

Effective Communication Plan: A robust communication strategy is vital for sharing quality-related information.

Regular meetings and updates can address quality concerns promptly. For example, scheduling weekly meetings to review quality metrics and discuss any issues.

Documentation of Discussions: Keeping detailed records of meetings can serve as a reference for future decisions and help track progress on quality issues.

Documentation and Record Keeping:

Comprehensive Documentation: Maintaining records of inspections, test results, and corrective actions tracks quality performance and provides evidence of compliance. For instance, documenting every inspection carried out on-site to ensure accountability.

Audit Trails: Organised documentation is essential for regulatory reviews and audits. For example, having a well-maintained log of all materials used, including their certifications and testing results, can help during compliance checks.

Identifying Quality Risks

Proactively identifying quality risks is essential for effective project planning. Understanding potential quality issues allows teams to implement strategies to mitigate risks.

Historical Data Review: Reviewing past project data can highlight common quality risks. For example, if previous projects faced issues with moisture intrusion, implementing better waterproofing measures in the current project would be wise.

Industry Trends Analysis: Staying informed about changes in industry standards and new technologies helps identify potential risks. For instance, adopting modern construction methods or materials can mitigate risks associated with traditional practices.

Training and Adaptation: Encouraging continuous education for team members on the latest industry practices can help them better manage risks. For example, providing training on new software for project management can enhance efficiency and accuracy.

Collaborative Risk Assessment

Multidisciplinary Approach: Involving diverse team members in the risk assessment process ensures thorough evaluation. For instance, a workshop that includes architects and engineers can lead to a comprehensive understanding of potential structural risks.

Risk Workshops: Conducting workshops to brainstorm potential risks and their impacts allows teams to identify issues before they escalate.

Contingency Planning

Developing Contingency Plans: Creating contingency plans for unavoidable risks is crucial. For example, allocating extra budget for unexpected repairs or setting aside additional time for potential delays.

Regular Review of Plans: Continuously updating contingency plans based on ongoing evaluations helps keep them relevant.

Checklist for Quality Management Plan

To ensure a thorough Quality Management Plan is in place before construction begins, use the following checklist:

Set specific, measurable quality objectives that align with client expectations and industry standards.

Develop a comprehensive Quality Management Plan outlining roles, responsibilities, and communication protocols.

Establish quantifiable quality metrics to measure project success at various stages. Conduct a thorough risk assessment to identify potential quality risks and develop mitigation strategies.

Ensure all team members are trained in quality control processes, industry standards, and project requirements.

Schedule regular reviews and updates of the Quality Management Plan throughout the project lifecycle to address changes or challenges.

Establish clear quality objectives and developing a detailed management plan, construction firms can ensure their projects meet high-quality standards from start to finish. This structured approach enhances the quality of work and builds trust with clients, stakeholders, and regulatory bodies, contributing to the long-term success of the project. This commitment to quality control is crucial in creating safe, durable, and aesthetically pleasing structures.

■■

CHAPTER 4

SELECTING AND QUALIFYING CONTRACTORS

A skilled and reliable contractor ensures that work meets quality standards, stays on schedule, and adheres to safety regulations, while poor selection can lead to delays, cost overruns, and compromised quality.

Implementing a Pre-Qualification Process

Choosing the right contractors is critical to maintaining quality throughout the construction process. A pre-qualification process helps to filter out contractors that lack the necessary experience, expertise, or quality control practices. This process should be thorough and objective, considering factors beyond cost, such as:

Experience with Similar Projects: Contractors who have successfully completed projects similar to yours will be more familiar with the unique challenges and quality requirements. This is especially important in luxury construction, where the level of detail and finish is significantly higher.

Financial Health: Financially stable contractors are less likely to face disruptions due to cash flow problems. Assessing a contractor's financial statements can help predict their ability to complete the project on time and without sacrificing quality.

Capacity and Resources: Contractors must have the manpower, equipment, and technical expertise necessary to meet project demands. This includes having specialised skills or access to subcontractors who can handle specific high-quality finishes or technical installations.

Criteria for Contractor Evaluation and Selection

The evaluation process should not focus solely on price but also consider the contractor's overall capability to deliver a high-quality result. Evaluation criteria should include:

Past Project Performance: Review case studies or site visits of the contractor's past work to gauge their ability to meet quality and time objectives. Luxury construction projects in particular demand close attention to detail, and past performance is an excellent indicator of future success.

Client References: Contacting previous clients provides insights into the contractor's reliability, communication skills, and ability to handle complex quality demands. Ask about their experience with the contractor's adherence to quality standards and any challenges faced during the project.

Quality Management Systems: Contractors should have established quality control practices, including regular site inspections, detailed workmanship standards, and a process for addressing non-compliance. Look for contractors with ISO 9001 certification or an equivalent quality management system that ensures continuous monitoring of quality.

Training Programs for Workers: Assess whether the contractor provides ongoing training for its workers and subcontractors. Contractors who invest in training programs typically have better workmanship and are more adept at adhering to stringent quality control standards.

Kaizen Practices for Contractor Performance

Incorporating Kaizen into contractor performance evaluation can lead to continuous improvement in quality management. Encourage contractors to adopt Kaizen principles, where they regularly assess and refine their construction processes. Continuous feedback loops and incremental improvements can significantly enhance project outcomes.

Checklist for Contractor Pre-Qualification and Selection

Review the contractor's past performance and project history, focusing on similar projects.

Assess financial stability through detailed financial reviews.

Ensure the contractor has the necessary capacity, resources, and skills.

Contact past clients to gather feedback on performance and quality.

Evaluate the contractor's quality management system and certifications.

Check for continuous training programs for workers and adherence to quality control practices.

Encourage contractors to implement Kaizen practices for continuous improvement.

By following a structured pre-qualification and evaluation process, construction firms can ensure they are partnering with contractors who will uphold the highest standards of quality and workmanship throughout the project

CHAPTER 5

EFFECTIVE SITE MANAGEMENT

Effective site management coordinates resources and maintains communication to ensure construction activities meet quality standards through regular inspections and risk mitigation

Conducting Site Inspections

Site inspections are one of the most crucial components of quality control in construction. They ensure that every stage of the project is completed in alignment with the project's design, safety protocols, and workmanship standards. To maximise the effectiveness of inspections, they must be systematic, frequent, and targeted. Key areas of focus during site inspections include:

Compliance with Design Specifications:

Inspectors must ensure that construction activities strictly follow the architectural and engineering plans. Any deviation from the design can compromise the structural integrity, aesthetic appeal, or functionality of the final product. During inspections, materials should be verified against specifications, and construction techniques should be assessed for compliance with detailed drawings and plans.

Safety Practices:

Every construction site is a dynamic environment with inherent risks. Inspectors must ensure that all safety protocols are followed to protect workers and reduce hazards. This includes checking whether the site is equipped with proper personal protective equipment (PPE), that safety signage is visible and clear, and that all equipment and scaffolding are safely used and maintained.

Quality of Workmanship:

Inspecting workmanship is critical to ensure long-term performance and client satisfaction. Inspectors should evaluate whether the quality of the work meets established standards and industry best practices. Any substandard work should be corrected immediately to prevent rework or future defects. Real-time feedback during inspections helps workers address any deviations from expected quality, ensuring that issues are corrected before they compound.

Expanded Role of Inspections

In luxury construction projects, attention to detail is paramount. Beyond the structural requirements, inspectors must also assess the aesthetics, alignment, and finishing details of materials such as custom wood, marble, or glass. Inspections should include checking for uniformity in materials, accurate alignment of structures, and consistency in finishing. The higher the level of detail required, the more frequent and focused these inspections need to be.

Implementing Quality Control Checks

A comprehensive schedule for quality control checks is essential to ensure quality at every stage of the project. These checks should be planned according to key milestones and project phases:

Pre-construction Checks: Ensures all materials are available, properly stored, and meet the quality requirements before work begins.

Ongoing Quality Checks: Conducted during critical construction phases, such as pouring concrete, erecting steel frameworks, or installing HVAC systems. These checks ensure that the work aligns with quality objectives and prevents potential defects.

Final Quality Inspections: Performed before project completion, focusing on final finishes, installation of fixtures, and the overall presentation of the work. This step ensures that the project is completed to the highest standard before handover.

In addition to scheduled checks, random spot checks are useful for identifying problems that may not be caught during routine inspections. Random checks increase accountability, as workers are aware that any phase of the project can be inspected at any time. They also ensure that quality is maintained across all areas, not just the ones subject to planned inspections.

Establishing Corrective Action Protocols

Even with rigorous quality control processes, issues may arise. Having a clear corrective action protocol ensures that issues are identified, addressed, and prevented from recurring. This process includes:

Documentation of Issues: Every quality-related issue should be logged in detail. This log should include a description of the concerns, its location, the materials involved, and the personnel responsible for the affected work. Accurate documentation allows for transparency and accountability.

Root Cause Analysis

Identifying the root cause of quality failures is essential for preventing them from happening again. Whether it's a materials defect, improper techniques, or a communication breakdown, understanding the underlying problem allows for targeted corrective actions.

Implementing Corrective Measures: Once the root cause is identified, develop a strategy to resolve the issue. This may involve reworking the affected area, replacing faulty materials, or adjusting workflows. Implementing corrective actions promptly ensures that the issue doesn't spread to other parts of the project.

Lessons Learned: Use the findings from the root cause analysis to improve future projects. Integrating these lessons into the firm's quality management system ensures continuous improvement and the prevention of recurring issues.

In-House Audit Method

In-house audits are a proactive approach to maintaining quality throughout the construction process. Conducting weekly audits ensures compliance with quality control standards and provides opportunities to address small issues before they become larger issues.

Weekly Site Audits: A designated team should conduct audits focusing on quality control processes, ensuring that the project adheres to the defined standards. These audits should be documented and followed up with reports on any non-conformances.

Standardised Audit Checklist: Use a checklist that includes key quality parameters such as material storage, installation techniques, safety compliance, and workmanship. This standardised approach ensures consistency across audits and provides a comprehensive overview of the project's quality performance.

Documentation of Findings and Corrective Actions

Audit Findings: Clearly documenting audit findings is the first step in addressing quality concerns. This includes a detailed description of the identified issue, supported by evidence such as records, data, or observations that substantiate the finding.

Corrective Action Plan: Once the issue is identified, a corrective action plan should be developed. This plan includes both immediate actions to resolve the problem and long-term strategies to address the root cause and prevent future occurrences. Each action should have an assigned responsible party and a defined timeline for completion.

Tracking and Follow-up: Tracking the implementation of corrective actions is critical to ensuring they are carried out as planned. Regular monitoring should be conducted, and the effectiveness of the actions must be verified to confirm that the issue is fully resolved. Documentation of completion ensures that all actions are properly logged and reviewed.

Reporting: Finally, a comprehensive audit report should be compiled, summarising the findings, corrective actions, and any preventive measures taken. This report should be shared with relevant stakeholders to maintain transparency and accountability.

By maintaining thorough documentation and following a structured approach to corrective actions, organisations can resolve quality issues efficiently and prevent future occurrences.

CHAPTER 6

ENSURING QUALITY WORKMANSHIP

"Ensuring quality workmanship is not just about meeting standards—it's about embedding precision, skill, and pride into every task, turning a structure into a legacy."

Developing Training and Certification Programs

Quality workmanship starts with a skilled and knowledgeable workforce. Investing in comprehensive training and certification programs for all personnel involved in the construction process is key to ensuring consistent, high-quality output. Training programs should be tailored to specific roles and cover:

Quality Standards and Best Practices: General training on the quality standards required for the project, including material handling, safety protocols, and construction techniques.

Role-Specific Skills: Specialised training for workers in specific trades, such as plumbing, electrical work, masonry, or finishing. Certification programs give workers confidence in their skills and ensure that they meet industry benchmarks.

Ongoing Education and Certifications: Encourage continuous learning through additional certifications in new technologies, materials, or construction methods. This helps to keep the workforce current and competitive in a fast-evolving industry.

Defining Workmanship Standards

Workmanship standards must be clearly defined to ensure consistent quality across all trades. Standards should cover all aspects of construction, from material handling to final finishes, and should be detailed enough to provide clear guidance.

Materials Handling: Proper handling and storage of materials are critical to maintaining their quality. Guidelines should include the correct methods for storing materials such as wood, steel, and concrete to prevent degradation or damage.

Construction Techniques: Best practices for executing construction tasks, such as laying foundations, erecting walls, and installing systems, must be outlined in detail. This ensures quality, safety, and efficiency across the project.

Final Finishing: Aesthetic quality is especially important in luxury construction. Workmanship standards should include detailed guidelines for final finishing work, such as plastering, painting, tile installation, and cabinetry, ensuring that the end product meets the highest aesthetic standards.

Inspection and Testing Methodologies

Regular inspections and testing are critical to assessing workmanship quality as the project progresses. These methodologies should be rigorous and well-documented to ensure they provide actionable feedback.

Routine Inspections: Conduct daily or weekly inspections to monitor ongoing work. These inspections should focus on critical tasks that are prone to defects or errors, such as electrical wiring, plumbing, and concrete pouring. Inspectors should provide immediate feedback to workers to resolve any issues.

Performance Testing: For structural and system components, performance tests should be conducted to verify compliance with design specifications. Examples include:

Load Testing: For structural elements like beams and columns.

Pressure Testing: For plumbing and HVAC systems to check for leaks or weaknesses.

Documentation of Findings: Keep detailed records of all inspections and testing. These records provide a clear audit trail, ensuring accountability and facilitating continuous improvement.

Kaizen Practices

Incorporating Kaizen principles fosters a culture of continuous improvement, ensuring that workers and supervisors are constantly investigating how to enhance quality. Key aspects of Kaizen practices include:

Workshops: Regular workshops where team members can discuss current practices and suggest improvements. Workers are closest to the construction process and can often identify inefficiencies or quality risks that supervisors may miss.

Encouraging Feedback: Create channels for workers to provide feedback on quality control processes. Continuous feedback loops help address quality issues quickly and improve the overall workflow.

Celebrating Small Wins: Recognise and celebrate minor improvements and successes, which builds a positive culture focused on quality. This could be as simple as rewarding teams for successfully implementing a new quality control measure.

Training Program for Engineers and Workers

Training programs for engineers and workers should be ongoing and comprehensive, focusing on both technical skills and quality management:

Introduction to Quality Control: Provide an overview of quality control principles, emphasising their importance in ensuring the success of the project and safety on the construction site.

Workmanship Standards Training: Hands-on workshops that allow workers to practice and refine their skills in specific areas of construction, such as carpentry, masonry, or HVAC installation.

Safety Protocols: Regular safety training sessions, focusing on minimising risks on-site and ensuring that all workers understand and adhere to the latest safety regulations.

■■

CHAPTER 7

COMPREHENSIVE TESTING AND INSPECTION

In construction, testing, and inspection are essential to ensure that all components of a project meet the required quality, safety, and performance standards. A well-structured testing and inspection framework enables project managers to identify issues early, make necessary adjustments, and prevent future problems. This chapter focuses on the various types of tests, the scheduling and frequency of inspections, and the importance of documentation and reporting in quality control.

Overview of Testing Types and Procedures

The construction process involves multiple systems, materials, and structural elements, each requiring specific types of testing to verify compliance with design specifications and safety standards.

Understanding the purpose of each test is crucial for comprehensive quality assurance.

Material Testing: This type of testing evaluates the properties and performance of construction materials before they are used. Some critical material tests include:

Compression Testing (for concrete): Ensures the concrete's strength meets design specifications, which is critical for load-bearing components like foundations, columns, and slabs.

Tensile Testing (for steel): Measures the steel's ability to resist tension, which is essential for reinforcing bars and structural steel.

Moisture Content Testing (for wood): Determines the moisture levels in wood to prevent warping or decay, especially in projects involving fine woodworking and custom finishes.

Durability and Weather Resistance Tests (for materials like glass or coatings): Ensures materials withstand environmental stresses over time.

Structural Testing: This includes tests that assess the load-bearing capacity and structural integrity of the building's core elements:

Load Tests: Verifies that floors, beams, and other structural elements can handle the anticipated live and dead loads. It ensures that safety margins are met for critical structural elements.

Settlement Monitoring (for foundations): Tracks any movement of the foundation over time to identify potential issues like uneven settling, which could compromise the structure.

Mechanical, Electrical, and Plumbing (MEP) Systems Testing: These systems are vital for building functionality and must be rigorously tested for safety and performance:

Pressure Testing (for plumbing): Ensures that the system can withstand the pressure of water or air without leaking. This is vital to prevent water damage and ensure system longevity.

Electrical Circuit Testing: Confirms that electrical installations meet design requirements and are safe for use. This includes load tests to verify circuits are not overloaded and grounding tests to ensure safety.

HVAC Performance Testing: Verifies that heating, ventilation, and air conditioning systems function efficiently and meet design performance criteria, such as maintaining proper temperature, air quality, and humidity levels.

Fire and Safety Systems Testing: Critical for life safety, these systems must be thoroughly tested:

Sprinkler System Testing: Ensures sprinklers will activate in the event of a fire.

Smoke and Fire Alarm Testing: Verifies that all alarms are functional and can alert occupants in emergencies.

Each type of testing requires specific procedures, tools, and personnel. A detailed testing plan should be developed during the pre-construction phase, specifying who is responsible for each test, when it will be conducted, and how results will be documented.

Scheduling and Frequency of Tests

Testing must be integrated into the overall construction timeline. The frequency and timing of tests depend on project complexity, the materials used, and the specific risks associated with various construction phases. Scheduling tests too late may result in costly rework, while testing too early can overlook issues that arise later in the process.

Pre-Installation Testing: Conducted before materials are used on-site, these tests ensure that only compliant materials are installed. For example, steel bars should undergo tensile testing before being embedded in concrete, while windows and glass should be tested for durability before installation.

During-Construction Testing: Key structural elements should be tested during critical construction phases. For example, load tests on concrete beams or columns should occur after they've cured but before additional construction layers are added. Similarly, electrical and plumbing systems should be tested before walls are closed to ensure any issues can be easily addressed.

Post-Construction Testing: After the completion of major construction activities, final tests must be conducted to confirm that all systems are fully functional and meet performance criteria. These include HVAC balancing tests, final load tests for structural integrity, and fire safety system checks.

Regular interval testing should be scheduled based on the project's progress. For example, materials like concrete may require testing at specific intervals to assess their curing progress and final strength.

Importance of Real-Time Inspections

In addition to scheduled testing, real-time inspections are vital for ensuring ongoing compliance with quality control standards. These inspections allow project managers and quality control personnel to identify potential problems before they escalate into costly issues. Real-time inspections should focus on the following areas:

Workmanship Quality: Regular checks ensure that workers are following established best practices and that their work meets design and quality standards.

Safety Protocols: Inspections should also confirm that safety regulations are being followed to prevent accidents or hazards that could compromise the project.

Compliance with Specifications: Inspectors should regularly review the construction process to confirm that all activities are in line with design specifications and industry regulations.

Documentation and Reporting Processes

Effective documentation and reporting are crucial for maintaining transparency, tracking progress, and ensuring accountability throughout the construction process. Every test and inspection conducted must be properly recorded, including detailed results and any corrective actions taken. This documentation serves several important purposes:

Creating an Audit Trail: Documenting all tests and inspections provides a clear record of the project's quality control measures. This is essential for regulatory compliance, client assurance, and future project reviews.

Tracking Corrective Actions: If a test or inspection identifies a problem, the corrective action taken must be recorded. This ensures that the issue is resolved effectively and provides insights for preventing similar problems in the future.

Generating Reports for Stakeholders: Regular reports should be generated for key stakeholders, including project managers, clients, and regulatory bodies. These reports should summarise the results of recent tests, highlight any issues, and outline plans for addressing deficiencies.

Centralised Documentation Systems should be used to store and organise all records. Modern project management software often includes modules for quality control that can automate the process of generating reports and tracking test results.

Continuous Feedback Loop

An important component of any quality control system is the feedback loop, where the results of tests and inspections are used to inform ongoing construction efforts. By analysing test outcomes and inspection reports, project managers can make data-driven decisions to enhance quality throughout the project's lifecycle.

Learning from Issues: When issues arise, the feedback loop allows teams to learn from the experience and adjust their approach. This could involve revising work methods, improving material selection, or refining testing protocols.

Preventing Future Failures: The feedback loop helps prevent future quality failures by continuously improving processes. When a root cause is identified, lessons learned are shared across teams to ensure similar issues don't occur on future projects.

Checklist for Testing and Inspection Procedures

A comprehensive checklist helps ensure that all critical aspects of testing and inspection are covered. This checklist can be adapted for the specific needs of each project but should include the following general elements:

Test Scheduling:

Ensure all required tests are scheduled at the appropriate stages of construction.

Verify that tests align with project milestones to avoid delays.

Pre-Test Preparation:

Confirm that testing personnel have the necessary tools, equipment, and documentation.

Ensure that materials and systems to be tested are ready for evaluation.

Conducting Tests:

Follow testing protocols rigorously, ensuring that the tests are carried out under the specified conditions.

Record all test data accurately and immediately.

Post-Test Documentation:

Log all test results into the project's centralised database.

Document any failures or issues and record the corresponding corrective actions.

Reporting Findings:

Generate reports summarising test results, highlighting any discrepancies, and detailing follow-up actions.

Share reports with relevant stakeholders to ensure transparency and accountability.

Follow-Up Actions:

Verify that any corrective actions are implemented correctly and promptly.

Re-test as necessary to confirm the issue has been fully resolved.

Effective testing and inspection are the cornerstones of quality control in construction. By employing a comprehensive approach that includes rigorous material and system testing, real-time inspections, detailed documentation, and a continuous feedback loop, construction firms can ensure that projects meet the highest standards of quality and performance. The ultimate goal is not just to meet regulatory requirements but to exceed client expectations and deliver structures that stand the test of time.

CHAPTER 8

STRUCTURAL QUALITY CONTROL

Structural quality control is a critical element of construction management, focusing on ensuring the integrity and durability of the building's load-bearing elements. Key components of structural quality control involve rigorous testing and inspection of materials like concrete and steel, as these are the foundation of any structure's strength. This chapter will delve into the essential techniques, procedures, and measures used to maintain high standards in structural quality control.

Concrete Testing and Inspection Techniques

Concrete is one of the most widely used materials in construction, forming the foundation, columns, beams, and slabs of most buildings. Ensuring its quality involves rigorous testing and inspections throughout the construction process.

Slump Tests

One of the most basic yet crucial tests for fresh concrete is the slump test, which measures the workability and consistency of the concrete mix. Workability directly affects how well the concrete can be placed and compacted, impacting the overall strength and durability.

Procedure: The slump test is performed by filling a cone-shaped mould with fresh concrete, then lifting the mould to allow the concrete to slump. The decrease in height is measured to assess workability.

Importance: A concrete mix that is too wet can lead to segregation of materials, while a mix that is too dry may be difficult to work with, leading to voids or incomplete compaction.

Compressive Strength Tests:

The compressive strength of concrete is its ability to resist axial loads without crushing, a critical factor in structural stability. Strength tests should be conducted at various stages of curing to ensure the concrete reaches its specified design strength.

Procedure: Samples of concrete, usually cast into cylinders, are cured under controlled conditions and tested at specific intervals—typically at 7, 14, and 28 days—to determine the progressive development of strength.

Importance: Ensuring that the concrete meets or exceeds its specified compressive strength is essential for the safety of load-bearing structures. Early identification of low-strength concrete can help implement corrective actions before further work progresses.

Temperature Monitoring:

Proper curing of concrete is temperature-dependent. Too much heat during the hydration process can lead to thermal cracking, compromising the structural integrity. Conversely, low temperatures can slow down the curing process, resulting in weak concrete.

Monitoring Techniques: Employ temperature sensors embedded in concrete to track internal heat development during curing. Adjustments such as cooling methods or protective coverings can mitigate extreme temperature effects.

Importance: Monitoring temperature ensures that the concrete cures uniformly, avoiding cracks and weaknesses that could cause structural failure over time.

Non-Destructive Testing (NDT)

Non-destructive techniques such as ultrasonic pulse velocity or rebound hammer testing are increasingly used to assess the quality of hardened concrete without damaging it.

Ultrasonic Testing: Measures the time it takes for an ultrasonic pulse to travel through concrete, providing insights into its homogeneity and strength.

Rebound Hammer Test: A surface test that gives a relative measure of surface hardness, indicating possible defects or weak spots.

Steel Quality Assurance Measures

Steel is another critical structural material, used in reinforcing concrete, constructing frames, and other structural elements. Ensuring its quality is vital for preventing failures related to load-bearing capacity and overall structural performance.

Visual Inspections:

Visual inspections are the first line of defence in steel quality assurance. These inspections focus on identifying visible defects such as ***corrosion, rust, pitting, or cracks*** that could compromise the structural performance of steel.

Procedure: Conduct regular inspections both before and after the steel is delivered to the site, and again after it has been installed. Inspect welds, connections, and surfaces for signs of deterioration or damage.

Importance: Early identification of physical defects prevents the use of substandard steel that could jeopardise the project's safety.

Ultrasonic Testing:

Ultrasonic testing (UT) is a more advanced inspection technique used to detect internal flaws in steel that are not visible to the naked eye. It is especially critical for evaluating the integrity of welds and ensuring there are no voids or cracks within the material.

Procedure: A probe emitting ultrasonic waves is placed on the surface of the steel. These waves penetrate the material, and any reflections from internal flaws are detected by the equipment.

Importance: Ensuring that welds and steel members are free from internal defects is critical for preventing structural failures, particularly in high-stress areas like beams and columns.

Material Certifications:

Every batch of steel used in construction should be accompanied by proper ***certification documents*** that verify its composition and mechanical properties, such as tensile strength, ductility, and yield strength. These certifications ensure the material complies with industry standards and project specifications.

Procedure: Verify material certificates from suppliers and cross-check them with design specifications.

Importance: Using certified materials guarantees that the steel will perform as required under various loads and environmental conditions.

Load Testing for Steel Structures:

After installation, steel components may be subject to load testing to confirm that they can support the expected loads without deformation or failure.

Procedure: Simulate loading conditions by applying distributed or concentrated loads to beams, columns, or other structural members.

Importance: Load testing verifies that the installed steel can handle real-world stresses, providing reassurance of structural integrity.

Checklist for Concrete and Steel Testing:

Concrete Testing:

Conduct slump tests on-site before pouring concrete.

Schedule compressive strength tests at 7, 14, and 28 days.

Monitor curing conditions using temperature sensors to prevent thermal cracking.

Use non-destructive testing methods like ultrasonic or rebound hammer tests to assess hardened concrete.

Steel Quality Assurance:

Perform visual inspections of steel for rust, cracks, or surface defects.

Use ultrasonic testing to inspect welds and detect internal flaws.

Verify that all steel used is accompanied by appropriate material certifications.

Conduct load tests on critical steel members to confirm performance stressed.

"Structural quality control is the backbone of safe construction. It ensures that every component meets design specifications, withstands loads, and prevents future failures. Without it, even the best designs are at risk. Safety starts with structural integrity."

■■

CHAPTER 9

QUALITY CONTROL IN MEP SYSTEMS

Mechanical, electrical, and plumbing (MEP) systems form the vital infrastructure of any building, providing essential services such as heating, cooling, electricity, water supply, and waste management. The successful integration of these systems is fundamental to ensuring that buildings operate efficiently, safely, and comfortably. However, the complexity and interconnectivity of MEP systems make quality control indispensable to overall building performance. This chapter outlines in detail the rigorous testing, inspection, and commissioning procedures needed to ensure the functionality, compliance, and longevity of MEP systems, safeguarding both the structure and its occupants.

Testing and Inspection of MEP Systems

Ensuring that MEP systems function correctly and meet design specifications requires a combination of pre-installation checks, during-construction inspections, and post-installation testing. These measures ensure functionality and long-term reliability. Proper coordination among the mechanical, electrical, and plumbing trades is necessary to prevent system conflicts or failures, which can arise from improper installations or material choices.

Leak Testing (Plumbing Systems)

Leakage in plumbing systems can lead to extensive damage, mould growth, and long-term structural issues if undetected before components are sealed behind walls or floors. Therefore, rigorous leak tests are vital during installation to ensure the system is watertight and ready for use.

Procedure: The system should be pressurised with water or air, and pressure levels should be carefully monitored over a set period to detect any drops that would indicate a leak. A thorough visual inspection of all joints, seals, and fixtures must accompany the pressure testing to verify the integrity of each component.

Equipment:

Pressure Test Pumps: Used to pressurise plumbing systems with water.

Compressed Air Test Kits: For air pressure testing.

Manometers/Pressure Gauges: For monitoring pressure levels.

Leak Detectors (Electronic or Dye Kits): Used for detecting leaks at joints and seals.

Importance: Detecting leaks early on allows for quick repairs, reducing the risk of costly future repairs, extensive water damage, and delays to the project schedule.

Load Testing (Electrical Systems)

Electrical systems must be thoroughly tested under operational load conditions to verify their ability to handle the intended power demands of the building without failures such as tripped breakers, overloaded circuits, or system blackouts.

Procedure: Connect the electrical systems to their intended loads and simulate real-world operating conditions. Monitor system performance, voltage drops, and temperature variations during testing. Ensure both the main and backup power supplies are tested under these conditions.

Equipment:

Load Banks: Simulate electrical loads for testing power supply capacity.

Digital Multimeters: Measure voltage, current, and resistance.

Thermal Imaging Cameras: Detect overheating or potential failures in electrical systems.

Power Quality Analysers: Monitor power fluctuations, harmonic distortions, and voltage drops.

Importance: Load testing is critical for ensuring safety and reliability. Systems that cannot handle the expected power demand could result in frequent outages, equipment failure, or even fire hazards, posing significant risks to both the building and its occupants.

Functional Testing (Mechanical Systems)

Mechanical systems, especially heating, ventilation, and air conditioning (HVAC), must undergo comprehensive functional testing to verify that they meet design performance criteria, ensure occupant comfort, and comply with energy efficiency standards.

Procedure: Operate the HVAC systems under different environmental conditions and assess various parameters such as airflow, temperature control, humidity, and system efficiency. This may involve simulating extreme weather conditions or varying building loads to ensure optimal performance.

Equipment:

Anemometers: Measure airflow rates in ducts and outlets.

Thermometers and Hygrometers: For measuring temperature and humidity levels.

Smoke Pens or Fog Machines: To assess air distribution and identify leaks in ventilation systems.

Balometer : Used to measure and balance air volume flow at diffusers and grilles.

Energy Meters: Track the efficiency of HVAC systems by measuring power consumption.

Importance: Proper functional testing guarantees that HVAC systems operate as designed, providing a comfortable indoor environment, maintaining energy efficiency, and adhering to building codes. Unchecked issues can lead to poor indoor air quality, discomfort, or excessive energy consumption.

Effective Commissioning Procedures

Commissioning is the final step in verifying that MEP systems meet the project's design intent and are fully operational. This process is essential for identifying any last-minute discrepancies and ensuring all systems are functioning together harmoniously.

System Integration Tests

MEP systems rarely operate in isolation; they are interdependent, and any failure in one system can directly affect another. Integration testing ensures that MEP systems, such as HVAC, plumbing, and electrical, work together seamlessly to provide optimal building performance.

Procedure: Simulate real-world operational conditions, including load fluctuations and stress tests, to observe how the systems interact. For example, monitor how the HVAC system responds when electrical demand increases, or assess whether plumbing systems maintain functionality under varying water pressures.

Equipment:

Data Loggers: For tracking system performance over time.

Building Management System (BMS) Software: Used to monitor and integrate all MEP systems.

Flow Meters: Monitor water flow rates during integration testing.

Temperature and Pressure Sensors: For tracking system responses during operation.

Importance: System integration tests are critical for identifying potential bottlenecks, inefficiencies, or failures that could arise during actual building operations. These tests ensure that systems operate cohesively and efficiently.

Documentation of Results

Proper documentation of commissioning activities is essential for tracking compliance with design and performance standards, facilitating future maintenance, and providing an official audit trail.

Procedure: Document all test results, including any deviations from expected outcomes, and the corrective actions taken to resolve issues. Maintain a centralised database for easy access to this documentation.

Equipment:

Digital Recording Tools: Tablets or laptops with specialised software for organising test data and generating reports.

Cameras: To capture visual evidence of system performance or failures.

Importance: Comprehensive documentation is a key part of quality assurance. It ensures that the building systems are functioning correctly at the time of commissioning and serves as a reference for future system modifications, troubleshooting, or upgrades.

Post-Commissioning Support

Post-commissioning support is essential for maintaining the performance and reliability of MEP systems. Building systems evolve, and ongoing monitoring and maintenance ensure that they continue to function efficiently and safely.

Procedure: Develop a structured maintenance plan that includes periodic system checks, troubleshooting guides, and updates to account for changing building operations. Regular check-ins with building operators help identify any emerging issues early.

Equipment:

Diagnostic Tools: For assessing system performance during routine checks.

Maintenance Scheduling Software:

To track maintenance schedules and system updates.

Importance: Ongoing support ensures that MEP systems continue to operate as intended. Preventive maintenance reduces the risk of system failure, improves system longevity, and can lead to cost savings by avoiding major repairs.

In-House Audit Method for MEP Systems

A dedicated in-house audit team ensures continuous quality control for MEP systems throughout both the construction and operational phases. This team plays a pivotal role in ensuring the systems meet both design specifications and safety regulations.

Establishing an Audit Team

The audit team should consist of specialists in mechanical, electrical, and plumbing systems. Their responsibilities include reviewing system installations, conducting spot inspections, and ensuring all systems meet design and safety standards.

Using a Checklist for Audits

A detailed checklist should guide every audit, ensuring no aspect of the MEP systems is overlooked. The checklist should include:

Verification of material quality and certifications.

Inspection of system installations for compliance with design drawings.

Confirmation that all pressure, load, and functional tests have been performed.

Documenting and Addressing Issues

When discrepancies or issues are identified, they must be documented and resolved quickly to avoid delays or future problems.

Procedure: Record issues in a structured tracking system, assign responsibility for corrective actions, and ensure re-inspections are completed to verify that problems have been addressed.

The structural integrity and functional performance of any building depend on its structure and also on its MEP systems. Thorough testing, inspection, commissioning, and ongoing auditing, with the support of appropriate equipment, are critical for ensuring these systems perform reliably and efficiently. Quality control is an ongoing process that does not end with construction completion but continues throughout the building's lifecycle.

■■

CHAPTER 10

SUSTAINABILITY IN QUALITY CONTROL

Sustainability has evolved into a critical cornerstone of modern construction practices, increasingly intertwined with quality control measures. Integrating sustainable methods ensures compliance with progressive regulations and enhances the overall value, durability, and long-term environmental impact of construction projects. In this chapter, we delve into how sustainability intersects with quality control, focusing on sustainable material selection, waste reduction strategies, and comprehensive environmental impact assessments (EIAs). By adopting these practices, construction projects can significantly reduce their ecological footprint, contribute to environmental preservation, and promote sustainable growth in the built environment.

Using Sustainable Materials and Practices

The construction industry is a major contributor to environmental degradation, with material consumption, energy use, and waste generation having a substantial impact. Consequently, implementing sustainable practices becomes essential for minimising environmental harm while upholding high standards of quality. One of the foremost ways to achieve this is by carefully selecting materials and employing eco-friendly construction techniques that align with modern sustainability goals.

1. Material Selection

Prioritise Sustainable Materials: A key step toward reducing the environmental footprint of a project is the careful selection of materials. Opting for materials like recycled steel, reclaimed wood, and low-VOC (Volatile Organic Compound) paints minimises environmental harm while promoting healthier indoor air quality. These materials meet stringent quality control criteria and enhance the overall sustainability of the project.

Certifications and Standards: To ensure materials meet high environmental standards, prioritise those that carry certifications such as LEED (Leadership in Energy and Environmental Design), BREEAM (Building Research Establishment Environmental Assessment Method), or Cradle to Cradle certification. These internationally recognised standards evaluate the environmental impact of materials throughout their lifecycle, from production to disposal.

Local Sourcing: Choosing locally sourced materials can further minimise the carbon emissions associated with transportation, contributing to a project's sustainability. Additionally, local materials are often better suited to the regional climate and environmental conditions, improving the overall durability and performance of the structure.

2. Waste Reduction Strategies

Construction Waste Management: Construction generates significant waste, but much of it can be minimised with strategic planning. Implement a robust waste management plan that outlines recycling, reuse, and responsible disposal methods. For example, concrete can be recycled and reused in future projects, while scrap metals can be repurposed, significantly reducing the waste burden on landfills.

Design for Deconstruction (DfD): Designing with future disassembly in mind is another effective strategy for reducing construction waste. DfD principles enable the reuse of materials when a building is decommissioned, reducing the need for new raw materials and preventing demolition waste.

Lean Construction Principles: Lean construction focuses on maximising value and minimising waste. This involves optimising resource use, preventing over-ordering materials, and streamlining workflows to eliminate inefficiencies, rework, and unnecessary delays.

Conducting Environmental Impact Assessments (EIA)

Environmental Impact Assessments (EIAs) are crucial in modern construction, serving as a strategic tool to identify and mitigate the potential environmental risks of a project. Conducting EIAs ensures that quality control measures extend beyond technical and material standards to encompass environmental considerations, safeguarding the ecosystem, and adhering to environmental regulations.

1. Evaluating Potential Impacts

Air Quality: Construction activities, including excavation, demolition, and heavy equipment use, can contribute to poor air quality through dust, emissions, and other pollutants. Quality control measures should include air quality assessments and mitigation strategies, such as dust suppression techniques and emissions controls, to minimise harm to workers and the surrounding community.

Water Resources: Construction can have a significant impact on water resources, including contamination through runoff and the improper disposal of hazardous materials. EIAs should identify these risks and incorporate water management plans, such as erosion control and stormwater treatment systems, to protect local water supplies.

Local Ecosystems: Construction near sensitive habitats can disrupt ecosystems, threatening biodiversity. Environmental impact assessments should evaluate potential threats to wildlife, vegetation, and soil quality, implementing measures such as habitat restoration, buffer zones, or conservation efforts to mitigate these effects.

2. Mitigation Strategies

Pollution Prevention Plans: Developing and executing pollution prevention plans that address key environmental concerns like dust control, water management, and waste disposal is essential for minimising a project's ecological footprint. These strategies prevent environmental harm and ensure compliance with environmental laws and community health standards.

Sustainable Site Management: Utilising eco-friendly equipment, stabilising soil, and preserving existing vegetation are all effective methods of sustainable site management. These practices contribute to the long-term environmental health of the site and promote sustainability within the project's lifecycle.

Energy Efficiency: Incorporating energy-efficient practices such as using renewable energy sources, optimising equipment usage, and reducing on-site energy consumption can greatly reduce greenhouse gas emissions and cut operational costs. Sustainable energy management is integral to achieving both quality control and sustainability goals.

3. Ongoing Monitoring

Environmental Monitoring Protocols: Continuously monitoring environmental indicators such as air and water quality, noise levels, and waste generation throughout the construction process ensures that sustainability goals are being met. These protocols allow for real-time adjustments and help maintain regulatory compliance.

Adaptive Management: Construction projects should be flexible in their approach to sustainability, using data from ongoing monitoring to adapt practices as needed. This approach helps mitigate any unforeseen environmental impacts, ensuring that sustainability measures are consistently effective.

Checklist for Environmental Assessments

Planning Stage:

Assess the project's potential environmental impacts on air quality, water resources, and local ecosystems.

Identify high-risk areas and develop mitigation strategies.

Ensure alignment with national and local environmental regulations.

Material Selection:

Select low-impact materials and prioritise those with sustainability certifications.

Favour locally sourced and recycled materials where feasible.

Waste Management:

Implement a waste reduction plan with recycling and reuse strategies.

Track and monitor waste generation throughout the project.

Mitigation Strategies:

Develop pollution prevention and site management plans.

Incorporate energy-efficient techniques to minimise resource consumption.

Ongoing Monitoring:

Establish continuous monitoring of environmental impacts.

Adapt construction practices based on real-time environmental data.

By integrating these sustainable practices—ranging from eco-friendly material selection and waste reduction strategies to thorough EIAs—quality control measures are aligned with technical and regulatory standards and contribute to long-term environmental stewardship. Construction projects that prioritise sustainability are better equipped to meet future challenges, reduce their environmental footprint, and set a standard for responsible development in the industry.

■■

CHAPTER 11

ESSENTIAL QUALITY CONTROL DOCUMENTATION

In the construction industry, effective documentation forms the backbone of a robust quality control system. The accuracy and completeness of records ensure that quality-related activities are properly tracked, analysed, and continuously improved upon. Thorough documentation satisfies regulatory and contractual requirements and enhances project transparency, accountability, and communication among stakeholders. As a tool for continuous improvement, well-maintained records allow project teams to learn from experiences, mitigate risks, and improve the overall quality of current and future projects.

In this chapter, we explore the various types of essential quality control documentation, best practices for maintaining record-keeping standards, and the implementation of document control systems to ensure efficient management of information throughout a project's lifecycle.

The Importance of Comprehensive Record-Keeping

Comprehensive record-keeping in quality control is essential for several reasons. Firstly, it ensures compliance with legal and regulatory standards, protecting the project and the organisation from potential litigation or disputes. Secondly, it provides a clear and accurate history of a project's quality control activities, which can be referenced for future projects or used to identify and address recurring issues. Finally, maintaining

proper documentation promotes transparency, holding all parties accountable for their roles in ensuring quality.

1. Inspection Reports

Inspection reports are vital for monitoring quality throughout the construction process. These reports provide a clear record of the findings from site inspections, allowing project managers and stakeholders to address any issues promptly.

Detailed Records

Every inspection report should include detailed information, such as the date of the inspection, the personnel involved, the specific areas inspected, the findings, and any corrective actions required or taken. Detailed records ensure that nothing is overlooked and that issues are addressed comprehensively. By systematically capturing all relevant details, reports serve as both a historical reference and a valuable tool for continuous quality improvement.

Utilisation of Digital Tools

The integration of digital platforms in construction has significantly improved the process of maintaining inspection reports. Digital tools and mobile applications enable inspectors to document findings directly from the job site. This real-time data entry reduces the risk of errors, ensures that information is captured promptly, and allows for faster communication of critical issues. Digital records also make it easier to share information with stakeholders and create a seamless chain of documentation from inspection to corrective action.

Review and Sign-off Procedures: A review and sign-off process ensures that inspection reports are verified by qualified personnel. After an inspection, reports should be reviewed by a supervisor or quality manager, who can validate the findings and confirm the adequacy of any proposed corrective actions. This adds an extra layer of scrutiny to the process and reinforces the importance of accuracy in quality control documentation.

2. Test Results

Testing is a crucial component of quality control, as it provides objective data about the performance and suitability of materials, components, and systems used in construction. Documenting test results comprehensively ensures that all stakeholders have access to verifiable information about the quality of the project.

Comprehensive Documentation

Test result documentation should not only include pass/fail outcomes but also detailed information about the methodologies used, the sample sizes, and the conditions under which the tests were conducted. This level of detail provides context for the results and ensures that future analyses can account for variables that may have affected the test outcomes.

Accessibility and Traceability

It's essential that test results be stored in a centralised and easily accessible database. This ensures that key stakeholders can quickly locate and review test data whenever necessary. Traceability is particularly important when discrepancies arise; being able to reference historical test data helps identify the source of problems and ensures transparency throughout the project.

Historical Comparison

Including historical test data in reports allows project teams to identify trends, compare current performance with past benchmarks, and detect patterns that may indicate emerging issues. For example, if the strength of concrete consistently improves with each test cycle, that might signal the effectiveness of a new material or process. Conversely, if there's a decline, it might highlight the need for immediate corrective action.

3. Change Orders and Correspondence

In the dynamic environment of construction, changes are inevitable. Documenting every change order and associated correspondence is crucial for maintaining quality control and ensuring that these changes do not negatively impact the project's quality, timeline, or budget.

Comprehensive Log Maintenance

Maintaining a comprehensive log of all change orders—including the reasons for the change, the associated costs, and approvals from relevant stakeholders—is essential for tracking adjustments made during the project. This log allows for a clear understanding of the scope changes and ensures that all parties remain aligned on expectations and deliverables.

Effective Communication

Every discussion related to quality control, whether it takes place via email, in-person meetings, or memos, should be thoroughly documented. These records prevent miscommunication and ensure that all decisions and the rationale behind them are clearly communicated to everyone involved. This practice reduces misunderstandings and provides a reference for future discussions.

Impact Assessment

Every change order should be accompanied by an impact assessment that evaluates how the change will affect the project's quality, schedule, and budget. Assessing these potential impacts and documenting them ensures that changes are aligned with the project's goals and do not compromise quality.

Implementing Document Control Systems

An effective document control system is necessary to manage the vast amount of quality control documentation generated during a construction project. These systems help ensure that all documents are accurately tracked, easily accessible, and regularly updated.

1. Version Control

Version control is a critical component of document management. Without it, teams risk working with outdated or inaccurate information, which can lead to significant quality issues.

Tracking Changes

Every document should have a unique identifier and version number to track changes over time. When a document is updated, a log should record the date, author, and the specific changes made. This process ensures that all stakeholders are aware of the most current version of the document and can access previous versions if needed for reference.

Clear Labelling

Consistent and clear labelling practices should be implemented across all project documentation. This ensures that team members can quickly identify the most current documents and prevents confusion that could arise from using outdated versions.

2. Distribution Protocols

Ensuring that all relevant stakeholders have access to quality control documents is essential for maintaining effective communication and collaboration.

Stakeholder Access: Establish clear protocols for distributing documentation to stakeholders, ensuring that the right people have access to the right information at the right time. This helps avoid delays and misunderstandings, ensuring that everyone is informed of key decisions and developments.

Regular Updates: Schedule regular document updates and notify stakeholders when significant changes occur. This ensures that quality control documentation remains current and that everyone involved in the project is aware of the latest developments. Regular updates also foster proactive management of quality issues, as stakeholders can quickly react to new information.

3. Retention Policies

Retention policies ensure that quality control documents are kept for the appropriate length of time to comply with legal and regulatory requirements, as well as for internal reference and future audits.

Defining Retention Periods: Clear retention policies should define how long quality control documents need to be kept and when they can be disposed of. These policies should comply with local laws and industry regulations, ensuring that all necessary records are available for the required duration.

Secure Storage Solutions: Secure storage, especially for sensitive documentation, is crucial. Digital storage solutions with access controls help protect documents from unauthorised access, while ensuring that authorised personnel can easily retrieve necessary information. Secure storage also safeguards documentation against loss or damage, ensuring that records remain intact for future reference.

Checklist for Effective Record-Keeping

Complete Inspection Reports: Ensure that all inspection reports are completed and filed accurately, with sufficient detail to serve as a historical record and tool for future analysis.

Maintain Testing Logs: Keep detailed logs of testing procedures and results, ensuring consistency in the methodologies used and accessibility for future reference.

Track Non-Conformance Reports: Systematically track and document non-conformance reports identifying recurring issues and areas for improvement.

Update Documentation Protocols: Regularly review and update documentation protocols to reflect the best practices and the evolving needs of the project.

Provide Training on Record-Keeping: Conduct training sessions for all project personnel on the importance of effective record-keeping, reinforcing the role documentation plays in maintaining project quality.

Comprehensive documentation is at the heart of effective quality control. By establishing rigorous record-keeping standards and implementing reliable document control systems, project teams can ensure transparency, accountability, and continuous improvement throughout the project lifecycle. This systematic approach to documentation ensures compliance with regulatory standards and promotes a culture of excellence, where every stakeholder is committed to maintaining and improving quality at every stage of the project.

■■

CHAPTER 12

COMPREHENSIVE REPORTING ON QUALITY CONTROL

Effective reporting on quality control activities is integral to the success of construction projects. Reporting serves as the primary communication tool for stakeholders, ensuring that quality standards are met, discrepancies are addressed, and continuous improvement efforts are fostered throughout the project lifecycle. In addition to maintaining transparency, systematic reporting helps in identifying potential risks early and supports proactive decision-making. A robust quality control reporting framework facilitates structured communication and documentation, which in turn promotes accountability and enhances the overall performance of the project.

This chapter delves into the importance of standardised reporting formats, the establishment of clear frequencies for reporting, addressing discrepancies, and ensuring continuous follow-up. By examining these key components, we provide a roadmap for effective quality control reporting.

Establishing Standardised Reporting Formats and Frequencies

To ensure consistency and clarity in communication, it is essential to adopt standardised formats for reporting on quality control activities. These reports should provide comprehensive data on the quality performance of the project while highlighting actionable insights that inform decision-making. Each report should be tailored to suit the needs of its audience, from project managers to contractors, ensuring the right level of detail is provided to all stakeholders.

1. Summary of Findings

Comprehensive Overview: Each quality control report should begin with a detailed summary of findings from inspections, testing, and other quality control activities. This summary should highlight key observations, potential risks, and trends identified over time. Visual elements such as graphs, charts, and heat maps can be invaluable tools for illustrating data, making it easier for stakeholders to understand complex information and make informed decisions.

Focus on Actionable Insights: Reports should focus on documenting quality issues and offering actionable insights. Highlighting areas that require immediate attention helps prioritise corrective actions and allocate resources effectively. By directing attention to critical areas, the report facilitates proactive decision-making, preventing minor issues from becoming major problems.

2. Action Items and Corrective Actions

Clear Outline of Corrective Actions: Every report should clearly outline the corrective actions required to resolve the quality issues identified during inspections or testing. It is critical that these corrective actions are precise, with clear timelines, assigned responsibilities, and any necessary resources detailed. This structured approach ensures that everyone involved in the project understands what is required and by when.

Follow-Up Mechanisms: Establishing a robust system to track the implementation and completion of corrective actions is crucial for maintaining accountability. Regular follow-up should be integrated into team meetings to assess progress, discuss potential roadblocks, and adjust timelines if necessary. These review sessions help ensure that quality improvements are achieved on time and that all team members remain focused on quality goals.

3. Stakeholder Communication and Reporting Frequency

Timely Reporting: Regular reporting on quality control activities ensures that all stakeholders are updated on the project's progress. A consistent reporting schedule—whether weekly, bi-weekly, or monthly—should be established based on the project's needs and complexity. For more dynamic projects, more frequent reporting may be required to address quality concerns promptly.

Transparent Communication Channels: Leveraging centralised platforms for reporting and sharing information is vital for promoting transparency and collaboration. A centralised communication hub allows stakeholders to access up-to-date reports, discuss potential issues, and contribute to solutions, creating a more cohesive approach to quality management.

Addressing Discrepancies and Resolution Processes

Effectively addressing discrepancies is crucial for maintaining high-quality standards. Clear, well-defined processes for reporting discrepancies and resolving quality issues ensure that problems are identified, investigated, and corrected promptly. Establishing a framework for addressing discrepancies also fosters a culture of accountability, where all team members understand their roles in upholding quality standards.

1. Incident Reporting

Documenting Incidents: Establish procedures for reporting and documenting quality-related incidents as soon as they are identified. Documentation should include the nature of the incident, its potential impact on the project, and any immediate actions taken to contain the issue. This documentation is essential for immediate corrective action, future analysis and improvement.

Root Cause Analysis(RAC)

Once an incident is reported, conducting a thorough root cause analysis is essential for identifying the underlying factors that led to the discrepancy. This process should involve all relevant stakeholders and aim to develop actionable recommendations that address both the immediate issue and any systemic problems. Root cause analysis helps prevent similar incidents from recurring and ensures continuous improvement in quality management practices.

2. Investigation Protocols

Systematic Investigation: Establishing standardised protocols for investigating quality discrepancies ensures consistency and thoroughness. This process should include steps for gathering relevant data, analysing the situation, and involving appropriate stakeholders in the investigation. A systematic approach to investigations ensures that all issues are thoroughly examined, and the right corrective measures are implemented.

Corrective Action Planning: After an investigation is completed, develop a detailed corrective action plan. This plan should include specific timelines, designated personnel responsible for implementing the corrective actions, and any necessary resources. The plan should also account for potential barriers to implementation and offer contingency measures to ensure the successful resolution of the issue.

3. Follow-Up and Verification

Implementation of Corrective Actions: Ensuring that all corrective actions are effectively implemented is crucial for resolving quality issues. Follow-up responsibilities should be assigned to specific team members who will monitor progress and provide regular updates to the project team.

Verification through Inspections: After corrective actions have been implemented, conduct follow-up inspections to verify their effectiveness. These inspections ensure that the root causes of the discrepancies have been addressed and that the project is back on track to meet quality standards. All verification activities should be documented to maintain a complete and accurate quality control record.

Checklist for Reporting Quality Issues

Clear Protocols for Reporting Discrepancies: Establish and communicate clear protocols for reporting quality issues. All team members should understand how to report discrepancies and feel confident that their concerns will be addressed promptly and effectively.

Standardised Reporting Formats

Use standardised formats for reporting quality issues to ensure clarity and consistency. This practice helps streamline communication and ensures that all relevant information is included in reports.

Regular Review Meetings: Schedule regular meetings to review quality issues and their resolutions. These meetings foster a culture of continuous improvement, allowing the team to reflect on lessons learned and make proactive adjustments to future work.

Open Communication Channels: Encourage open communication regarding quality concerns. A proactive approach to problem-solving helps prevent minor issues from escalating and encourages all team members to take ownership of the project's quality.

Comprehensive Documentation

Document all discussions and decisions related to quality issues to maintain a complete record. This documentation is vital for both ongoing project management and future reference, providing valuable insights for improving quality control processes on subsequent projects.

Conclusion

By implementing standardised reporting and maintaining clear communication channels, construction projects can significantly enhance their quality control efforts. Effective reporting promotes accountability, encourages proactive problem-solving, and ensures that stakeholders remain informed throughout the project. Moreover, addressing discrepancies in a structured and timely manner prevents potential issues from becoming significant roadblocks and fosters a culture of transparency and continuous improvement.

Robust reporting practices help maintain compliance with industry standards and strengthen relationships among stakeholders, driving the success of the project as a whole. Establishing a well-defined reporting framework, complete with clear protocols for addressing quality issues, ensures that the entire project team remains focused on delivering high-quality outcomes, even in the face of challenges.

■■

CHAPTER 13

QUALITY CONTROL AUDITS AND ASSESSMENTS

Quality control audits are an indispensable component of construction management. They help ensure compliance with established standards and provide an essential framework for evaluating and enhancing quality control measures across various projects. Regular audits offer a structured approach to scrutinising processes, identifying areas for improvement, and instilling a culture of accountability among team members. This chapter explores the significance of both internal and external audits, effective strategies for assessment and improvement, and the implementation of in-house audit methodologies.

Conducting Internal and External Audits

Quality control audits serve as critical tools for assessing the performance and effectiveness of construction practices. By conducting regular audits, Organisations can gain valuable insights into their operations, highlighting strengths while pinpointing weaknesses that may require attention.

1. Importance of Regular Audits

Internal Audits: These audits are conducted within the organisation to assess compliance with established quality control standards and practices. Internal audits are instrumental in identifying areas of strength that can be leveraged and weaknesses that necessitate corrective action. They encourage a culture of self-assessment and continuous improvement, allowing teams to take ownership of their processes. Furthermore, these audits can boost team morale as employees recognise that their contributions are vital to maintaining quality.

External Audits: Engaging external auditors provides an objective and independent review of quality control measures. These audits can uncover issues that may be overlooked internally and offer fresh perspectives on industry best practices. External auditors bring expertise and experience that can enhance the overall quality control framework, ensuring adherence to regulatory requirements and industry standards. Additionally, the involvement of external auditors can bolster client confidence in the project's quality and compliance.

Strategies for Assessment and Improvement

To drive continuous improvement based on audit findings, organisations must implement effective strategies that address identified issues and optimise quality control processes.

1. Action Plans

Developing Action Plans: After audits are completed, it is essential to develop actionable plans that address the issues identified. Each action plan should clearly outline the steps required to rectify the problem, assign responsibilities to specific team members, and establish timelines for completion. This structured approach ensures accountability and promotes timely resolution of quality issues. It is also beneficial to prioritise action items based on their impact and urgency, allowing teams to focus on critical areas first.

Follow-Up Mechanisms: Regularly review the progress of action plans through follow-up meetings and status reports. This allows the team to adjust strategies as necessary and ensures that accountability is maintained throughout the process. Furthermore, establishing a routine for these follow-ups helps keep quality control at the forefront of project management.

2. Performance Metrics

Establishing Metrics: To evaluate the effectiveness of implemented changes, organisations should establish clear performance metrics. These metrics can include defect rates, inspection pass rates, and customer satisfaction scores, among others. By quantifying the impact of changes, teams can make informed decisions about future quality control initiatives. It may also be beneficial to incorporate both leading and lagging indicators, providing a comprehensive view of quality performance.

Benchmarking Against Best Practices: Use established industry benchmarks to compare performance metrics against peers. This analysis helps identify areas for improvement and sets realistic goals for quality control efforts. Engaging with industry groups to access benchmarking data can enhance this process and promote accountability across the board.

3. Feedback Mechanisms

Creating Feedback Channels: Establish feedback mechanisms that encourage input from team members regarding quality control processes. This could involve surveys, suggestion boxes, or regular meetings focused on quality improvement discussions. By fostering an open environment for feedback, organisations can gain valuable insights and promote a culture of continuous improvement. Additionally, utilising digital tools for feedback collection can streamline the process and make it easier for team members to share their thoughts.

Incorporating Feedback into Practices: Actively integrate feedback into quality control practices. When team members see that their input is valued and acted upon, it enhances engagement and commitment to quality initiatives. Moreover, acknowledging and responding to feedback publicly can further motivate team members to participate in the quality improvement process.

In-House Audit Method

Implementing a structured in-house audit method is essential for maintaining high-quality standards. This method involves engaging team members at all levels in the audit process to promote accountability and continuous improvement.

1. Scheduling Regular Internal Audits

Structured Audit Checklist: Develop a comprehensive checklist that covers all aspects of quality control measures. This checklist should include criteria for inspections, testing, documentation, and compliance with established standards. Regular audits should be scheduled to ensure consistency and thoroughness in evaluations. Utilising technology to automate checklist distribution and data collection can streamline the auditing process.

Timely Execution: Conduct audits at predetermined intervals—monthly, quarterly, or annually—depending on the project size and complexity. This regularity ensures that quality control measures are consistently monitored and improved upon. Consider integrating audits into regular project management schedules to ensure they do not become an afterthought.

2. Evaluating Effectiveness

Assessment Criteria: During audits, assess the effectiveness of quality control measures against established benchmarks. Identify areas of non-compliance or where improvements can be made, documenting findings for future reference. Ensure that audit findings are communicated transparently with all team members, fostering a sense of collective responsibility.

Actionable Insights: After each audit, compile findings into a report that highlights strengths and weaknesses. Share these insights with the team and use them as a basis for further training and development. These reports should focus on shortcomings and celebrate successes and progress made, fostering a balanced perspective.

3. Engaging All Team Members

Collaborative Approach: Involve all team members in the audit process, fostering a sense of ownership and accountability. Encourage team members to contribute their perspectives and insights during audits, promoting a collaborative approach to quality control. This involvement can take the form of participation in audit teams or being invited to present their observations during audit meetings.

Training and Development: Provide training on audit procedures and the importance of quality control. This investment in team development enhances their understanding and engagement with quality initiatives. Consider organising workshops or seminars that cover audit procedures and focus on broader quality control topics, equipping team members with a more comprehensive skill set.

Quality control audits and assessments are essential for maintaining high standards in construction projects. By conducting regular internal and external audits, organisations can gain valuable insights that drive continuous improvement. Implementing structured action plans, establishing performance metrics, and fostering open feedback channels all contribute to creating a robust quality control framework. In-house audit methods further engage team members in the process, promoting accountability and collective responsibility. Ultimately, these practices enhance the quality of construction projects and cultivate a culture of excellence and commitment among all team members. As organisations embrace these auditing strategies, they position themselves for greater success in an increasingly competitive landscape.

CHAPTER 14

BEST PRACTICES AND LESSONS LEARNED

In the construction industry, the continuous improvement of quality control processes is critical to delivering successful projects. By analysing real-world examples, setting benchmarks, and adapting to emerging trends, Organisations can ensure that they remain competitive and uphold high standards of quality. This chapter focuses on how Organisations can benefit from learning through case studies, establish industry benchmarks, explore future trends, and implement training programs for continuous improvement. These strategies help to enhance project outcomes and create a culture of quality and accountability.

Analysing Case Studies and Success Stories

Examining successful quality control implementations provides a wealth of knowledge and serves as a guide for refining future practices. Case studies offer concrete examples of strategies that have led to high-quality outcomes, allowing project teams to learn from real-life situations and avoid common pitfalls.

1. Identifying Success Factors

Case Study Selection: Selecting the right case studies is crucial to derive meaningful insights. Choose case studies that reflect diverse challenges and successes across different types of projects, regions, and market sectors. For example, case studies on large infrastructure projects may reveal the importance of communication across extended teams, while those on smaller residential projects may highlight the impact of hands-on leadership. Identify and document the success factors from these case studies, such as strong leadership, active stakeholder engagement, effective resource allocation, and a proactive approach to problem-solving.

Lessons Learned: From each case study, compile the key lessons learned. These lessons should focus on the actions, processes, and decisions that contributed to the project's success. For instance, early identification of risks and meticulous documentation may have led to more effective quality control management. Sharing these insights with other project teams can influence how they approach future challenges and enhance the overall quality control effort.

2. Continuous Learning

Knowledge Sharing: It is important to establish a system for storing and disseminating case studies and the associated lessons learned. Create a centralised repository or digital library where team members can easily access these case studies. Regularly review these lessons in team meetings or quality workshops, and encourage teams to actively discuss how they can apply these insights to their current projects. This promotes a culture of continuous learning and fosters the mindset that there is always room for improvement, regardless of past successes.

Setting Industry Benchmarks

Benchmarks are essential for measuring quality control performance and establishing improvement goals. Industry benchmarks are based on data and best practices and provide a standard for evaluating project performance across different aspects of quality management.

1. Developing Benchmarks

Collaborative Efforts: Collaborate with industry leaders, regulatory bodies, and professional associations to develop benchmarks that are relevant and comprehensive. By working closely with these stakeholders, Organisations can ensure that their quality control standards are aligned with current industry practices. This collaborative effort also provides the opportunity to gather input from multiple perspectives, ensuring that benchmarks reflect real-world challenges and solutions.

Data-Driven Insights: Use data from completed projects to establish benchmarks. Analysing project performance metrics such as defect rates, rework costs, inspection failure rates, and timelines for resolving quality issues can provide a clear picture of areas that need improvement. By identifying trends and recurring issues, Organisations can develop benchmarks that drive meaningful improvements in quality control.

2. Measuring Performance

Regular Evaluations: Once benchmarks are established, it is essential to regularly measure performance against them. This can be done through scheduled quality audits, inspections, and reviews of project performance metrics. The goal is to identify gaps between actual performance and benchmark standards, enabling project teams to focus on areas requiring improvement.

Recognition of Achievements: Recognising and celebrating when teams meet or exceed established benchmark goals is critical for maintaining motivation and fostering a commitment to quality. Rewarding high-performing teams or individuals with incentives, awards, or public recognition helps reinforce the importance of meeting quality standards and encourages others to strive for similar achievements.

Exploring Future Trends in Quality Control

The construction industry is evolving, with new technologies, methodologies, and regulatory requirements reshaping how projects are executed. Staying ahead of these trends is vital for maintaining high-quality standards and ensuring long-term success.

1. Technological Advancements

AI and Automation: Artificial intelligence (AI) and automation are rapidly transforming how quality control is managed in construction. AI-powered tools can analyse large datasets from past projects, identify patterns that suggest potential quality issues, and predict risks before they materialise. For instance, AI can be used to monitor sensor data on-site, predicting structural weaknesses or material defects. Automation tools can also streamline routine inspections and testing processes, reducing the likelihood of human error and ensuring more consistent quality across all project phases.

Digital Tools:

The adoption of digital tools such as construction management software, Building Information Modelling (BIM), and cloud-based collaboration platforms enables more efficient documentation, tracking, and communication throughout the project lifecycle. These platforms improve visibility into quality control activities, allowing for real-time updates, streamlined workflows, and enhanced collaboration among stakeholders. With digital tools, project teams can manage quality data more effectively and respond to issues more quickly.

2. Adapting to Regulatory Changes

Evolving Regulations: Construction regulations and industry standards are constantly evolving, and Organisations must stay updated to avoid penalties and ensure compliance. By closely monitoring changes in local, national, and international regulations, quality control teams can adapt their practices accordingly. This could involve updating quality control protocols, retraining staff, or adopting new technologies to meet stricter safety or environmental standards.

Client Expectations: In today's competitive market, clients are increasingly demanding higher-quality outcomes and more transparent processes. Engaging clients in discussions about quality expectations early in the project can help ensure that their requirements are fully understood and met. Incorporating client feedback throughout the project lifecycle enhances quality control and builds trust and loyalty, leading to long-term partnerships and repeat business.

Training Program for Continuous Improvement

A well-designed training program is essential for ensuring that project teams remain committed to quality control and are equipped with the skills needed to adapt to new challenges. By regularly training staff in quality management principles and fostering a collaborative environment, Organisations can drive continuous improvement.

1. Regular Training Sessions

Interactive and Tailored Training: Conduct regular, interactive training sessions that focus on continuous improvement principles such as Kaizen. These sessions should be designed to address the specific needs of the team, using real-world examples and hands-on exercises to ensure that the material is relevant and applicable to their daily work. Training should also cover the latest trends in quality control, including the use of AI, digital tools, and updated regulatory requirements.

2. Team Involvement

Collaborative Problem-Solving: Encourage team members to participate actively in brainstorming sessions focused on improving quality control processes. By involving the team in identifying and solving quality-related issues, Organisations foster a sense of ownership and commitment to continuous improvement. This collaborative approach ensures that everyone is invested in achieving quality goals, and it encourages the sharing of innovative ideas.

3. Sharing Lessons Learned

Knowledge Sharing Culture: Create opportunities for team members to share lessons learned from past projects. This could take the form of informal presentations, roundtable discussions, or written case studies. Sharing successes and challenges promotes a culture of openness and mutual learning, helping to spread best practices across the organisation and reinforcing the importance of quality control.

Conclusion

By analysing case studies, setting industry benchmarks, exploring future trends, and implementing continuous improvement training, Organisations can significantly enhance their quality control practices. This holistic approach fosters a culture of excellence, accountability, and collaboration, ensuring that construction projects should exceed quality expectations. The key to long-term success lies in continually learning from experiences, staying ahead of industry developments, and empowering teams to take ownership of the quality control process. Through these efforts, construction projects can achieve consistently high standards of quality, leading to stronger client relationships, reduced rework, and more successful outcomes.

CHAPTER 15

TOTAL QUALITY MANAGEMENT (TQM) IN CONSTRUCTION

Total Quality Management (TQM) is a management approach that focuses on continuous improvement and customer satisfaction throughout all organisational processes. In the construction industry, TQM serves as a strategic framework that enables organisations to enhance the quality of their projects, reduce waste, and improve overall efficiency. This chapter will delve into the principles of TQM, its implementation strategies in construction projects, the critical role of leadership and culture, and the measurable benefits it brings to organisations striving for excellence.

Principles of Total Quality Management

TQM is built upon several foundational principles that guide organisations in their pursuit of quality. Understanding and applying these principles is essential for effectively integrating TQM into construction projects.

1. Customer Focus

A customer-centric approach is the cornerstone of TQM. In construction, understanding client needs and expectations is paramount to delivering projects that meet or exceed these criteria. Key aspects include:

Identifying Customer Requirements: Organisations should conduct thorough discussions and surveys to ascertain client expectations regarding quality, timelines, and budget. This information serves as a foundation for all project planning and execution.

Engagement and Communication: Regular updates and open lines of communication with clients ensure transparency and trust. Engaging clients in decision-making processes can lead to better alignment of project goals and customer expectations.

Feedback Mechanisms: Establishing robust feedback channels allows clients to voice their opinions and experiences throughout the project. Utilising this feedback can lead to immediate adjustments, improving overall satisfaction and loyalty.

2. Continuous Improvement

Continuous improvement is integral to TQM, emphasising the need for Organisations to constantly assess and enhance their processes, products, and services. In the context of construction:

Kaizen Philosophy: Adopting the Japanese concept of Kaizen, or "continuous improvement," encourages small, incremental changes that lead to significant enhancements over time. This philosophy promotes a mindset where every employee is responsible for identifying areas for improvement.

Regular Training and Development: Continuous learning opportunities, including workshops and seminars on quality control techniques and best practices, ensure that employees are equipped to implement improvements effectively.

Process Review: Conducting periodic reviews of workflows and procedures enables Organisations to identify inefficiencies, redundancies, and bottlenecks that hinder quality.

3. Employee Involvement

TQM recognises that employee engagement is crucial for achieving quality objectives. In construction, this can be fostered through:

Empowerment: Employees at all levels should be encouraged to contribute ideas for improving quality and efficiency. Empowerment can lead to innovative solutions and a sense of ownership over the work performed.

Team-Based Approaches: Establishing cross-functional teams for problem-solving can enhance collaboration and leverage diverse perspectives. Team members should be involved in quality circles, where they regularly meet to discuss quality issues and propose solutions.

Recognition and Reward Systems: Implementing recognition programs for employees who contribute to quality improvement initiatives can motivate others to engage in TQM practices actively.

4. Process Approach

A process-oriented mindset is fundamental to TQM, focusing on understanding and improving processes to enhance quality in construction.

Mapping Processes

Creating process maps that outline workflows helps identify key steps and interdependencies. This clarity aids in spotting potential quality issues before they arise.

Standard Operating Procedures (SOPs)

Developing and adhering to SOPs ensures consistency in work execution. These guidelines should be regularly updated based on feedback and changing project requirements.

Root Cause Analysis

Employing tools like the Fishbone Diagram or 5 Why's technique helps identify the root causes of quality issues, enabling teams to address underlying problems rather than just treating symptoms.

5. Fact-Based Decision-Making

TQM promotes the use of data and analysis for informed decision-making. In construction, this involves:

Key Performance Indicators (KPIs)

Establishing KPIs such as defect rates, cost variance, schedule adherence, and client satisfaction scores provides a measurable framework for assessing performance.

Data Collection and Analysis

Organisations should implement systems for systematically collecting and analysing data related to quality performance. Statistical tools and software can help visualise trends and identify areas needing attention.

Benchmarking

Comparing performance against industry standards and best practices can help Organisations identify gaps and set realistic improvement goals.

6. Integrated System

TQM emphasises an integrated quality management system that encompasses all functions within an organisation. Key considerations include:

Cross-Department Collaboration

encouraging collaboration between different departments (e.g., design, procurement, and construction) ensures that quality standards are uniformly applied across the project lifecycle.

Quality Management Plans

Developing a comprehensive quality management plan that outlines objectives, processes, roles, and responsibilities ensures that everyone understands their contributions to quality.

Alignment with Organisational Goals

Quality objectives should align with the broader strategic goals of the organisation. This alignment helps prioritise quality initiatives and allocate resources effectively.

Implementing TQM in Construction Projects

The successful implementation of TQM in construction requires a structured approach. Below are the key steps Organisations can take to effectively adopt TQM principles

1. Establish a Quality Policy

A quality policy serves as a guiding document that outlines an organisation's commitment to quality. This policy should:

Define Quality Objectives: Clearly articulate the quality goals the organisation aims to achieve.

Communicate Commitment: Ensure that all employees understand the organisation's dedication to quality and their roles in achieving quality objectives.

2. Train Employees

Training is essential for fostering a culture of quality. Effective training programs should include:

Comprehensive Quality Education: Cover TQM principles, quality control techniques, and industry-specific standards.

Hands-On Training: Provide practical training opportunities that allow employees to apply quality concepts in real-world scenarios.

Regular Refreshers: Offer ongoing training sessions to keep employees updated on the latest quality practices and tools.

3. Set Measurable Objectives

Establishing clear and measurable quality objectives is crucial for tracking progress and driving continuous improvement. Organisations should:

Use SMART Criteria: Ensure that quality objectives are Specific, Measurable, Achievable, Relevant, and Time-bound.

Align Objectives with KPIs:

Connect quality objectives to specific KPIs to facilitate performance tracking.

4. Foster a Collaborative Culture

Creating a collaborative environment encourages communication and teamwork. Strategies for fostering collaboration include:

Regular Meetings: Hold meetings focused on quality discussions, allowing team members to share insights and experiences.

Cross-Functional Teams: Form cross-departmental teams to address specific quality issues or improvement initiatives.

Open Communication Channels: Establish platforms for employees to share ideas and feedback regarding quality practices.

5. Implement Quality Control Measures

Quality control measures are essential for ensuring compliance with established standards. Organisations should:

Develop a Quality Control Plan: Outline specific procedures for monitoring quality at each stage of construction.

Conduct Regular Inspections: Schedule inspections and testing throughout the project lifecycle to identify potential issues before they escalate.

Use Technology:

Leverage technology such as drones, Building Information Modelling (BIM), and quality management software to enhance monitoring and documentation processes.

6. Continuously Review and Improve

TQM is an ongoing process that requires regular evaluation and refinement. Organisations should:

Conduct Internal Audits: Schedule periodic audits to assess the effectiveness of quality management practices and identify areas for improvement.

Gather Feedback: Collect feedback from employees, clients, and stakeholders to inform quality improvement initiatives.

Implement Corrective Actions: Develop action plans to address identified issues and monitor the effectiveness of implemented changes.

Leadership and Organisational Culture

The success of TQM implementation heavily relies on strong leadership and a supportive organisational culture. Leaders play a pivotal role in fostering a quality-driven environment by:

Modelling Commitment: Leaders should demonstrate a strong commitment to quality through their actions and decisions, setting an example for employees.

Encouraging Innovation

Create an environment where employees feel safe to propose innovative solutions and challenge existing processes.

Promoting Accountability

Establish clear expectations for quality performance and hold team members accountable for their contributions.

Investing in Resources:

Allocate necessary resources, including training, technology, and personnel, to support quality initiatives.

Benefits of TQM in Construction

The adoption of TQM in construction offers numerous benefits, including:

1. Enhanced Quality

By prioritising quality in every aspect of a project, Organisations can deliver superior products and services that meet or exceed client expectations. This commitment to quality leads to fewer defects, rework, and delays, ultimately enhancing project outcomes.

2. Improved Client Satisfaction

A customer-focused approach fosters stronger relationships with clients. By actively engaging clients in the construction process and ensuring their needs are met, Organisations can enhance client satisfaction and build long-term partnerships.

3. Increased Efficiency

TQM promotes the identification and elimination of inefficiencies in construction processes. By streamlining workflows and improving communication, Organisations can reduce waste, minimise costs, and enhance overall project efficiency.

4. Greater Employee Engagement

Involving employees in quality initiatives fosters a sense of ownership and accountability. When team members feel valued and empowered, they are more likely to take pride in their work and contribute positively to the organisation's quality objectives.

5. Stronger Reputation

Organisations that consistently deliver high-quality projects gain a competitive advantage and establish a strong reputation in the industry. This reputation attracts new clients and enhances relationships with existing stakeholders.

6. Long-Term Sustainability

A robust TQM framework contributes to the long-term sustainability of construction firms. By embedding quality into their culture and operations, Organisations can build resilience against market fluctuations and maintain a steady stream of work.

Case Studies and Success Stories

To illustrate the effective application of TQM in construction, consider the following case studies:

Case Study 1: A Construction Firm's Journey to Excellence

A mid-sized construction firm faced challenges with project delays and client dissatisfaction due to quality issues. The firm decided to implement TQM principles across its operations. Key actions included:

Employee Training

All employees participated in a comprehensive training program focused on quality awareness and TQM principles.

Process Mapping : The firm mapped out its workflows and identified bottlenecks that contributed to delays.

Client Engagement

The firm actively sought client feedback throughout the construction process.

As a result, the organisation reduced rework by 30%, improved project delivery times, and received higher client satisfaction ratings. The firm's commitment to quality became a competitive advantage, leading to increased business and referrals.

Case Study 2: A Large Infrastructure Project

A major infrastructure project faced numerous challenges, including regulatory compliance issues and stakeholder dissatisfaction. The project team adopted a TQM approach to address these concerns:

Integrated Quality Management System

The team developed a comprehensive quality management plan that aligned with project objectives and stakeholder expectations.

Regular Audits and Reviews:

The team conducted regular internal audits to assess compliance with quality standards.

Collaboration with Stakeholders

Engaging stakeholders throughout the project lifecycle led to better alignment with their expectations.

As a result, the project was completed ahead of schedule, met all regulatory requirements, and received positive feedback from stakeholders. The successful implementation of TQM ensured project success and positioned the firm as a leader in quality construction.

Summary

Total Quality Management (TQM) is a powerful framework for enhancing quality in the construction industry. By focusing on customer satisfaction, continuous improvement, employee involvement, and process optimisation, Organisations can deliver superior projects that meet or exceed client expectations. The successful implementation of TQM requires a commitment to fostering a culture of quality, providing ongoing training, and integrating quality management practices into every aspect of operations.

As the construction industry continues to evolve, embracing TQM principles will enable Organisations to navigate challenges effectively and remain competitive in a dynamic marketplace. By prioritising quality and adopting a holistic approach to quality management, construction firms can achieve lasting success and build a reputation for excellence that sets them apart in the industry. The journey towards TQM may require time and effort, but the rewards—enhanced quality, satisfied clients, engaged employees, and a strong market position—are well worth it.

RECAP OF KEY TAKEAWAYS

1. Fundamental Principles of Quality Control

Quality control is not merely a collection of procedures; it is a comprehensive system designed to ensure that construction projects meet established standards and client expectations. The essential principles discussed throughout this book include:

Clear Standards and Specifications

Establishing clear and measurable standards and specifications is the cornerstone of effective quality control. These serve as benchmarks for evaluating all work, providing a foundation for consistency and reliability. Well-defined standards empower team members with concrete criteria to guide their tasks, ultimately enhancing overall productivity and project outcomes.

Systematic Inspections and Testing

Regular inspections and testing of materials, processes, and completed work are vital for early identification of issues and ensuring compliance with established standards. This proactive approach minimises the risk of defects and enhances overall project quality. By catching potential problems before they escalate, teams can save time and resources, contributing to a smoother project execution.

Comprehensive Documentation

Maintaining thorough documentation is critical for tracking quality control activities and providing a transparent record of compliance. This documentation includes everything from inspection reports to change orders, all of which contribute to accountability and informed decision-making. Proper documentation serves as a historical record and facilitates smoother transitions in project management, especially when onboarding new team members or stakeholders.

Continuous Improvement

The philosophy of continuous improvement is integral to quality control. By implementing feedback mechanisms and conducting regular audits, Organisations can identify strengths and weaknesses, fostering a culture of ongoing development and adaptation. This commitment to improvement encourages teams to remain agile and responsive to new challenges, ultimately elevating the quality of their work.

Total Quality Management (TQM)

TQM represents a holistic approach to quality management that emphasises customer satisfaction, continuous improvement, employee involvement, and process optimisation. By integrating TQM principles,

Organisations can cultivate a culture where quality is everyone's responsibility. This comprehensive methodology enhances project outcomes and strengthens the organisation's overall performance, adaptability, and competitiveness in the market.

2. Interconnectedness of Quality Control Processes

Throughout the book, we emphasised the interconnectedness of various quality control processes. From material selection and waste reduction strategies to internal audits and performance metrics, each aspect of quality control is interrelated. Understanding these connections enables teams to adopt a holistic approach, ensuring that quality is integrated into every phase of the construction process. This mindset enhances operational efficiency and leads to superior outcomes and client satisfaction. For example, a change in material selection can affect the overall project timeline, underscoring the need for cross-functional collaboration.

Importance of Ongoing Quality Control Efforts

Quality control should not be viewed as a one-time activity confined to the project's initial stages. Instead, it must be regarded as an ongoing commitment that evolves alongside industry standards and client expectations. The dynamic nature of the construction industry necessitates adaptable and responsive quality control efforts.

1. Adapting to Changing Industry Standards

The construction landscape is continuously evolving due to advancements in technology, regulatory changes, and shifting market demands. To remain competitive, Organisations must stay informed about these changes and incorporate them into their quality control practices. This involves regular training for team members, updating documentation protocols, and integrating new technologies that enhance quality assurance processes. For instance, Utilising Building Information Modelling (BIM) can improve coordination and accuracy, thereby reducing errors that might otherwise compromise quality.

2. Meeting Client Expectations

Today's clients are increasingly discerning, expecting high-quality workmanship, transparency, and responsiveness throughout the construction process. Ongoing quality control efforts ensure that projects remain aligned with client needs, fostering trust and long-term relationships. By actively engaging clients in discussions about quality standards and involving them in the decision-making process, construction firms can enhance client satisfaction and loyalty. Tailoring communication and reporting methods to meet client preferences can further strengthen these relationships, ensuring clients feel heard and valued.

3. Long-Term Success and Sustainability

A robust quality control framework contributes to the long-term success of construction firms. By prioritising quality, Organisations can reduce rework, minimise waste, and enhance overall project efficiency. This leads to cost savings and strengthens the firm's reputation in the industry. Incorporating sustainable practices into quality control efforts further positions Organisations as leaders in the field, appealing to environmentally conscious clients and stakeholders. Adopting green building practices and sourcing sustainable materials can distinguish a firm in a competitive market, showcasing its commitment to responsible construction.

4. A Culture of Excellence

Establishing a culture of excellence within an organisation requires an unwavering commitment to quality control. This culture fosters accountability and encourages team members to take pride in their work, resulting in higher motivation and engagement. When quality becomes a shared value within the organisation, it permeates every project, driving continuous improvement and ensuring outstanding outcomes. Recognising and rewarding quality-driven behaviours can reinforce this culture, motivating team members to uphold high standards and strive for excellence in their daily tasks.

In summary, the principles of quality control discussed in this book are vital for achieving high-quality construction outcomes, for ensuring the sustainability and success of construction firms in a competitive marketplace. By embracing ongoing quality control efforts and adopting a holistic approach, Organisations can navigate the complexities of the industry while consistently meeting and exceeding client expectations.

As the construction industry continues to evolve, Organisations that prioritise quality control will enhance operational efficiency while building lasting relationships with clients and stakeholders. By integrating the principles and practices outlined in this book into their everyday operations, firms can remain resilient, innovative, and responsive to the demands of an increasingly dynamic environment. Ultimately, a steadfast focus on quality is not just an operational necessity; it is a strategic advantage that can pave the way for enduring success.

Moreover, by embracing Total Quality Management (TQM) as a core component of their quality strategies, Organisations can amplify their commitment to quality, fostering a culture of continuous improvement and customer-centricity. This approach ensures that construction firms are well-equipped to adapt and thrive in a competitive landscape, enhancing their capacity for innovation and overall effectiveness in delivering exceptional projects.

APPENDICES

The appendices serve as invaluable resources aimed at deepening the understanding of quality control in the construction industry. They provide practical tools, clear definitions, and comprehensive references that industry professionals can utilise to implement effective quality control practices. Below is a detailed overview of each appendix, designed to facilitate easy access to crucial information.

Appendix A

Glossary of Quality Control Terminology

This section includes a comprehensive glossary of terms related to quality control in construction. It serves as an essential reference for industry professionals, ensuring clarity in the understanding of key concepts and terminology frequently used throughout the quality control process.

Key Terms Defined:

Quality Assurance (QA): A systematic process aimed at ensuring that construction meets specified standards before work commences.

Quality Control (QC): The operational techniques and activities utilised to fulfil requirements for quality.

Non-Conformance: A deviation from a specified requirement that must be addressed to maintain quality standards.

Kaizen: A Japanese term meaning "continuous improvement," referring to practices designed to enhance processes.

Inspection: The process of examining work to confirm it meets established standards.

Purpose and Usage: This glossary will aid both new and seasoned professionals in navigating the complexities of quality control. By providing clear definitions, it enhances communication among team members and stakeholders, fostering a shared understanding of quality-related discussions.

Appendix B

Sample Quality Control Checklists and Forms

This section provides various templates and examples of checklists and forms commonly used in quality control processes. These resources are designed to streamline quality assurance efforts, enabling construction teams to implement best practices effectively.

Checklists

Pre-Construction Quality Checklist: A comprehensive checklist to be utilised before the commencement of construction, ensuring that all necessary permits, plans, and materials are in place.

Daily Inspection Checklist: A form for documenting daily inspections of ongoing work, capturing key findings and any corrective actions required.

Materials Verification Checklist: A checklist for verifying the quality and compliance of materials delivered to the site, ensuring they meet project specifications.

Forms:

Non-Conformance Report (NCR) Form: A standardised form for documenting any deviations from quality standards, detailing the issue and corrective actions taken.

Corrective Action Plan (CAP) Template: A structured template for developing action plans in response to identified quality issues, outlining responsibilities and timelines for resolution.

Quality Control Audit Form: A form used during audits to evaluate compliance with quality control processes, facilitating structured assessments and improvements.

Implementation Guidance: Each template and checklist includes guidance on how to effectively utilise them in the field, along with tips for customisation to suit specific project needs. This practical approach empowers teams to adopt standardised practices that enhance quality control.

Appendix C

References to Relevant Industry Standards and Regulations

This section provides a curated list of key industry standards and regulations relevant to quality control in construction. This resource offers professionals a foundational understanding of the legal and procedural requirements that govern quality practices within the industry.

Key Standards

ISO 9001: An international standard that specifies requirements for a quality management system, applicable to any organisation seeking to improve quality and customer satisfaction.

ASTM Standards: Various standards set by ASTM International for materials testing and construction practices, ensuring quality and performance.

LEED Certification Standards: Guidelines for achieving Leadership in Energy and Environmental Design certification, focusing on sustainable building practices.

Regulatory Frameworks:

Occupational Safety and Health Administration (OSHA) Regulations: Standards for workplace safety that indirectly influence quality control by ensuring a safe working environment.

Local Building Codes: Regional codes and regulations that must be adhered to in all construction projects, emphasising compliance and quality assurance.

Resources for Further Study: This appendix includes references to online resources, publications, and professional Organisations that provide additional information and updates on quality control standards and regulations. These resources are crucial for professionals seeking to deepen their understanding and ensure compliance with industry expectations.

By Utilising the materials provided in these appendices, construction professionals can enhance their knowledge and application of quality control practices. These resources facilitate effective implementation and promote a culture of excellence and continuous improvement within the construction industry.

■■

OTHER BOOKS

By Author

Navigating India's Education Maze
A Comprehensive Guide to Engineering and
Medical Entrance Exams.

THE SOCIAL MEDIA
Generation and
Its Impact
on Society

www.ingramcontent.com/pod-product-compliance
Lightning Source LLC
Chambersburg PA
CBHW022157160726
48282CB00024B/221
* 9 7 8 9 3 3 4 3 2 8 2 2 6 *